'Through several fascinating Nordic case studies, this book calls attention to how the media industry needs to urgently get its own house in order and to play a major role in effective environmental management. This unique study provides an important contribution to an evolving area of environmental communication, while contesting the myth that it costs more to "go green" with regards to film production and distribution. As environmental protocols across the media industry need to keep evolving as the world tries to face up to the huge challenges of climate change, this book makes a good start in showing the way meaningful and sustainable changes can be made.'

— *Pat Brereton, Dublin City University, Ireland*

'As we become increasingly aware of the environmental impact of media and the potential damage the digital age wreaks upon the planet, it is incumbent upon us to map in detail the local effects of our insatiable demand for data in its myriad electronic forms. In an approach which mobilises important methodological innovations, Pietari Kääpä does just this for the Nordic world. His lucid prose captures the state of current debate and brings a fresh insight to this vital area of research.'

— *Gareth Stanton, Goldsmiths, University of London, UK*

Environmental Management of the Media

In recent years the widely held misconception of the media as an 'ephemeral' industry has been challenged by research on the industry's significant material footprint. Despite this material turn, no systematic study of this sector has been conducted in ways that consider the role of the media industries as consumers and users of a range of natural resources.

Filling this gap, *Environmental Management of the Media* discusses the environmental management of the media industries in the UK and the Nordic countries. These Nordic countries, both as a set of small nations and as a regional constellation, are frequently perceived as some of the 'greenest' in the world, yet, not only is the footprint of the media industries practically ignored in academic research, but the very real stakes of the industries' global impact are not comprehensively understood. Here, the author focuses on four key areas for investigating the material impact of Nordic media: (1) resources used for production and dissemination; (2) regulation of the media; (3) organizational management; and (4) labour practices. By adopting an interdisciplinary perspective that combines ecocritical analysis with interrogation of the political economy of the creative industries, Kääpä argues that taking the industries to task on their environmental footprint is a multilevel resource and organizational management issue that must be addressed more effectively in contemporary media studies.

This book will be of great interest to students and scholars of media, communication and environmental studies.

Pietari Kääpä is Associate Professor in Media and Communications at the University of Warwick, UK.

Routledge Studies in Environmental Communication and Media

Environmental Management of the Media
Policy, Industry, Practice

Pietari Kääpä

LONDON AND NEW YORK

from Routledge

First published 2018 by Routledge

2 Park Square, Milton Park, Abingdon, Oxfordshire OX14 4RN
52 Vanderbilt Avenue, New York, NY 10017

Routledge is an imprint of the Taylor & Francis Group, an informa business

First issued in paperback 2020

British Library Cataloguing in Publication Data
A catalogue record for this book is available from the British Library

Library of Congress Cataloging in Publication Data
Names: Kääpä, Pietari, 1977- author.
Title: Environmental management of the media : policy, industry, practice / Pietari Kääpä.
Description: Abingdon, Oxon ; New York, NY : Routledge, 2018. | Series: Routledge studies in environmental communication and media | Includes bibliographical references and index.
Identifiers: LCCN 2017060787 | ISBN 9781138649828 (hbk) | ISBN 9781315625690 (ebk)
Subjects: LCSH: Mass media--Management. | Environmental management. | Mass media--Economic aspects.
Classification: LCC P96.M34 K335 2018 | DDC 302.23/068--dc23
LC record available at https://lccn.loc.gov/2017060787

ISBN: 978-1-138-64982-8 (hbk)
ISBN: 978-0-367-45984-0 (pbk)

Typeset in Goudy
by Taylor & Francis Books

This book is, as always, for Yan.

Contents

Tables

Acknowledgements

I would like to thank the British Academy/Leverhulme for providing funding for my project from 2014 onwards. This support has facilitated numerous research trips to countries in Europe where I have had the pleasure of both accessing archives and talking directly with various governmental ministries, cultural institutions, environmental scientists, and media producers. Without this level of access, this book would simply not have been possible. The various individuals in these organizations gracefully devoted time out of their busy schedules to discuss their work, allowing me first-hand access to the complex machinations in integrating media policy with environmental concerns.

Various academic and professional colleagues have offered constructive feedback and all kinds of essential support on the project during its lengthy gestation, including Pat Brereton, Mette Hjort, Sean Cubitt, Stephen Rust, Salma Monani, Hunter Vaughan, Tommy Gustafsson, Birgit Heidsiek, Christian Toennesen, Richard Haynes, Philippa Lovatt, Jo Garde-Hansen, Dafydd Sills-Jones, and Hilmar Gudlaugsson.

1 Material implications of the media

Introduction

At the culmination of the latest round of negotiations in Paris in 2015 on global responses to climate change, the United Nations (UN) celebrated the acceptance of a global climate agreement with an unabashed sense of achievement. The participants were quick to draw attention to new developments such as enticing the previously reticent USA and China to fully ratify its calls. However, critics did not hesitate to point out that the agreement was conditional on a considerable degree of ambiguity and uncertainty, drawing attention to the extent of the commitments and the ability of the UN to enforce them. These critiques were shared internally within the UN as key participants such as Christina Figuoreira, the Chief Negotiator, expressed scepticism over the applicability of the agreement and the tangible plans proposed by many key participants to control their emissions. Simultaneously, while steps were taken to curtail heavy carbon industries and other significant emitters of greenhouse gases, these address only a fraction of global environmental hazards and provide no comprehensive methodologies for dealing with the wide scale of polluting elements and the factors that contribute to them. Policy decisions were understandably geared at curbing massive amounts of process and waste emissions on a global scale by targeting those industrial sectors with the largest footprint. Yet, many other key sectors were practically ignored in these discussions.

One such omission is the role of information and communication technologies (ICT). Greenpeace estimates that the sector consumed over 7 per cent of global electricity demand in 2012, projecting that this could exceed 12 per cent by 2017, and continue to grow at least 7 per cent annually through 2030, double the average rate of electricity growth globally (Greenpeace 2017). The global communications infrastructure powers the economy and increasingly shapes politics in most parts of the world. Furthermore, they facilitate the formation and consolidation of the interconnected 'scapes' of globalization that Appadurai (1996) discusses. It would not be an overstatement to suggest that ICT penetrates and consolidates modern social life. As media platforms and technologies reach ubiquity, they require a range of resources in ever-expanding amounts. Important legislation like RoHS (Restrictions on Hazardous Substances) and WEEE (Waste Energy and Electricity Emissions) regulate materials that go into the production and recycling of media devices. However, these regulations are often superseded by legislative loopholes and criminal undercurrents in the e-graveyards of the world.

Richard Maxwell and Toby Miller (2012) argue that ICT companies are very much complicit in the proliferation of electronic waste through circumventing legislation trying to curtail its harmful effects. Shortened life-spans as part of planned obsolescence strategies and the fashioning of media gear are just some of the complex considerations that influence the impact that the communications ecosystem has on the environment. In addition, some estimate that the internet consumes up to 8 per cent of the total electricity generation in the UK (Hodgson 2015). This is not even considering problems like print and the chemicals of the publishing industry and the substantial resource-intense usage in film and television production, all areas that require specific attention to their protocols and practice. Considering the environmental footprint of the ICT infrastructure, it is clear that more needs to be done outside of RoHS and WEEE. Why was this particular sector and industry not a target for the UN session? Why has there been so little intergovernmental discussion about the huge scale of the ICT industry?

Ecomedia

While the relationship between the media and the environment has met with substantial interest from academics, most studies to date have focused on representational concerns. They discuss the ways environmental considerations are conveyed and communicated on television (Good 2013), social networks (Anderson 2014; Parham 2015), journalism (Cox and Pezzullo 2014) or film (Ingram 2000; Brereton 2005; Willoquet-Maricondi 2010; Rust et al. 2014a; Kääpä 2014). Others have focused on the facilitation of public awareness and policy through representational means. Discussion often works on specific thematic concerns, such as climate change (Hibberd and Nguyen 2013) or green rhetoric in the press (Hansen 2011). Much of this work accepts the general view that the media industries contribute to societal awareness of environmental issues – they have a 'brainprint' that can shape individual behaviour and policy. These studies consolidate an ecocritical approach focusing on, for example, addressing how mainstream media normalizes perceptions of the human-nature dichotomy or how corporate communications use greenwashing rhetoric for public relations purposes.

While these works have advanced the field of environmental communications significantly on understanding the influence of the media on awareness of environmental considerations, they do not account for the wider socio-environmental or physical-material impact of the industries. In short, they rarely consider the resources that go into the infrastructure or the actual production mechanisms facilitating the conveyance of these messages. Sean

Cubitt's study *Ecomedia* (2005) was one of the first comprehensive studies to focus on these material concerns, outlining not only many of the materials that go into media production, exhibition, distribution and consumption but also initiating scholarly discussion on the environmental footprint of digital media. To these ends, Cubitt's work constructs a range of suggestions in drawing attention to the detrimental environmental effects of the media industry. Among these, the massive consumption of electricity by data server farms as well as the use of energy-intensive terminal devices form particular concerns when approaching digital communications from an environmental perspective. In addition to infrastructural concerns, the emergence of the social habits of a convergent media culture have an environmental significance not often addressed by policy or academic work. Pervasive or ubiquitous media indicates not only that we have to be constantly connected, or 'on' (Turkel 2008), but also necessitates that consumers worldwide keep consuming, proliferating media messages interconnected on multiple platforms, all requiring ever increasing amounts of resources to feed the habit.

Cubitt's work emphasizes the notion that those studying the media cannot only focus on texts but need to take into account the variety of contextual factors that underpin media production. Adrian Ivakhiv takes up this point by suggesting the focus of critical analysis would need to be on 'things, processes, and systems that support and enable the making and disseminating of cultural texts' (Ivakhiv 2007: 19). This approach underlines Cubitt's argument concerning the necessity to consider media in all its complex contextuality, meaning that the analyst must focus on both the resources and labour for media production in addition to the messages it carries. Ivakhiv considers this approach as a means to expand analytical frames outside the representational and the ideological:

> A holistic eco-cinecriticism closely analyzes not only the representations found in a film but the telling of the film itself – its discursive and narrative structures, its inter-textual relations with the larger world, its capacities for extending or transforming perception of the larger world – and the actual contexts and effects of the film and its technical and cultural apparatus in the larger world.
>
> (ibid.: 18)

These key works have paved the way for a material turn in our understanding of the nature of media. These include seminal works such as Lisa Parks' (2007) study of the material conditions of television networks, including cable and satellite television, and Jennifer Gabrys's *Digital Rubbish* (2011), focusing on the proliferation of digital media devices. At stake in these works is the realization that there is much more to an environmental understanding of the media than the focus on content and messages or even the waste and emissions generated by production activities. The shift in analytical scope urged by these works is on the considerable infrastructure facilitating communications, that is, on the parts that do not fit with a conventional instrumental understanding of the media as a means to facilitate human communications. Thus, Parks discusses both electronic waste and the electronic waves that enable communications while Gabrys explores the materials that go into the production of the devices used to consume and produce media, as well as their considerable afterlife. Similar calls are made by Jane Bennett's (2010) work on vibrant materialities and Jussi Parikka's on the geology of media (2015), which both draw from new materialism and media archaeology to shift focus to the matter that produces the media. In these perspectives, media is the means of communications premised on a range of material processes. These include geological, chemical, biological and electric processes that leave a material footprint that can be assessed. Positioning media as a process of substances and electronic impulses means that there is both more and less to media than anthropogenic cultural messages. The media are thus more than human and less than human and both sides must form a part of any analysis of its environmental impact.

Two recent collections have arguably consolidated this material turn in media studies. The first of these, Nicole Starosielski and Lisa Parks' *Signal Traffic* (2015), hosts articles on the digital infrastructure. The focus here is predominantly on the ICT infrastructure that shapes the conduits of modern society, including the material groundings, such as cable networks and energy infrastructures, essential for the communications industry. Janet Walker and Starosielski's *Sustainable Media* (2016) casts its scope even wider and covers most traditional and emerging areas of the media. Chapters focus on the different energies and practices that form the media we consume. Especially notable are Brennan's work on back-up cultures, which takes to task the use of 'immaterial' Cloud services and the proliferating file management habits it generates, and Vaughan's study of the resources used for the production of *Singing in the Rain* (1952). The latter provides the seeds for an approach he calls 'ecomaterialism' which shifts the critical attention to the environmental consequences of film production methods. These collections provide a wide-ranging consolidation of the argument for the necessity to understand media studies as a field firmly planted in the various materialities that comprise its very existence.

Meanwhile, other work has been instrumental in pushing the field in unexpected directions. Nadia Bozak's work (2012) is notable for providing an ecophilosophical exploration of the environmental roles of the cinematic image. She argues that the tangibility of films reroutes the spectator to consider the resources that went into the production of the text they witness and thus facilitates a heightened sense of awareness of their material conditions. Similar work is undertaken in Kaapa's (2011) collection on ecocinema audiences. Chapters consider the ways audiences take on board environmental messages and make use of them in their cognitive and ideological responses to climate change. Ecophilosophical work focusing on spectators and reception contexts is also present in collections such as *Chinese*

Ecocinema (Lu and Mi 2009) and *Transnational Ecocinema* (Gustafsson and Kaapa 2013). While the chapters contained in these studies focus mostly on philosophical aspects emerging from the texts studied, they also include work on the production apparatus of an ecological adventure like James Cameron's *Avatar* (2009) or on the uses of discarded DVDs as agricultural supplements.

Another strand of ecocritical work synergizes the political economy of the industry with environmental concerns. Maxwell and Miller's *Greening the Media* (2012) combines the infrastructural – both regulatory and especially the technological – scope of some of the earlier work with a focus on environmental accounting in media organizations. Through this, they explore the regulatory and financial motivations for introducing environmental considerations into the media industry. Maxwell and Miller's analysis suggests that the accountants of a media organization are best positioned to negotiate or minimize the footprint of the sector. They are key players in ensuring the industry is aware of the emissions it generates and developing practices to take stock of these responsibilities. Similar arguments have been made by Christian Fuchs (2008, 2014) in relation to exploitative labour conditions – both human and environmental – in the production of the digital media infrastructure. He argues that we need to consider environmental problems 'social problems, not technological' ones (Fuchs 2008: 308), as ultimately, environmental protocols are instituted through social and political negotiations. This is a significant intervention that will shape much of the approach of this book, especially in terms of understanding environmental concerns as management and regulatory practices.

The material turn

Despite all these recent works on material media, Walker and Starosielski argue that it 'is surprising how little ecocriticism has ... concentrated on the ecological impact of media-related things, processes, networks, systems, and infrastructures' (2016: 13). At stake here is a shift in the study of media from ideological representations to the processes that comprise the contextual existence of the industry. This requires both a rethink of the ontological potential of media as well as the epistemological foundations of media studies. The challenge to ontology comes from a shift in focus to media production and materiality as well as its impact on society. Rather than just understanding it as a communicator of environmental messages, it is taken as a material practice that has a footprint that can be traced and evaluated. The epistemological challenge comes from the need to rethink how the media has been studied. As we have suggested, the field to date has predominantly focused on brainprinting, but with this material turn it is increasingly clear that footprinting holds at least equal value as a topic of analysis.

This focus on the material qualities of the media provides an important intervention in exploring the physical repercussions of the industry. The mediacity of media has of course been a topic of debate at least since McLuhan's (1964) arguments on the media as a powerful force capable of changing social structures through technological evolutions. Thus, one could state that instead of the media being the message, it is now the material that is the message. Yet, we must be careful with the ways we incorporate materiality into our analytical approach so as not to allow our understanding of the media's environmental role be overtaken by technological or materialist determinism – that is, of a focus on materiality that undoes the many political and social influences that also play significant roles in these processes. Messages communicated by the media – both 'content' and internal communications about the industry's perceptions of its own environmental role – counteract and reinforce the contextual – the material – operations of their production and consumption. As we will see, discourse and textuality continue to be as relevant as ever and provide the means to understand how materialities influence the media and how media production and management shape the material base in return.

Using sustainability as an approach, Cubitt provides a productive explanation for the ways the balance between the material base of production and the anthropocentric management of the industry needs to be understood. This involves the need 'to make the media more committed to sustainability, to sustain the very media we use, and to make a world where media are sustenance' (Cubitt 2017: 14). According to this argument, all levels of the media ecosystem must be sustainable: the infrastructure for communications must adhere to ecological principles, the media must be able to generate enough economic and reputational capital to sustain itself, and finally, it needs to communicate messages of value to its audiences that, in turn, feed into the necessity for the sector to act on what it preaches.

Media management

As can be seen from this very brief outline of academic work on the subject, evaluating the environmental role of the media is a multilevel resource and organizational management consideration. Yet, much of the academic work to date tends to overlook the management of media organizations and production strategies as one of the key areas that needs to be analysed, focusing instead on areas such as the physical properties of materials or infrastructural considerations. Management in this context means organizational oversight of its operations, including accounting, but also how the development and implementation of such strategies link with wider regulatory and industrial developments in this sector. Instead of only conducting work on content or the physical propensities of the footprint, a study that combines ecocritical work with the study of operational and organizational areas of media companies is required.

For us, the field of media management provides this focus. For Albarran (2005), the field is concerned with the application of managerial and organizational analysis to the media industry. In practice, this concerns the ways the different operational levels of media companies are managed to meet regulatory requirements and economic considerations in a competitive multimedia environment. A key distinction is that media management, in this line of study, does not refer to the management of assets in software systems – which is how the industry sees it. Instead, it is the study of the conduct of contemporary media organizations in relation to human and natural resources, the logic of the economy, regulatory regimes, and societal and cultural values.

I will adopt key areas of this field to propose a combination of ecocritical analysis and media management. This approach forms the basis for what I call the environmental management of the media. The study of environmental media management is thus best characterized as an interdisciplinary combination of traditional media studies with media economics, corporate communications, politics and policy. To facilitate such an exploration, we need an integrated field of study that focuses on at least four key areas for investigating the material impact of the media:

Regulation of the industries

- domestic media regulations concerning the industry's environmental responsibilities (i.e. laws imposed by governmental ministries for environmental and cultural concerns, adoption of these into media practice);
- international frameworks of regulation (concerning issues such as energy usage, recycling standards, chemicals, metals, and trade in hardware);
- industrial affiliations with NGOs (cooperative partnerships with organizations like Julie's Bicycle or Greenpeace to promote sustainability and green effectiveness);
- self-regulation (establishment of often voluntary best practice protocols for specific media sectors, economic initiatives to subsidize the adoption of environmental strategies).

The organization of labour

- management strategies (the development and implementation of sustainability in an organization, top-down or bottom-up);
- employee protocols (organizational rules on implementing green directives, tactics on motivating employee sustainability);
- environmental HR (green runners, eco-supervisors and other roles designed to oversee sustainability in an organization);
- training workshops (coordination of sustainability training from external or intra-organizational sources).

Resources

- production pathways (the role of traditional and digital media technologies in media production; resourcing materials);
- carbon pricing (ways to improve or circumvent existing legislation by drawing on 'green taxation');

- digital data (awareness and regulation of the methods of digital data flow and storage);
- production protocols (procurement of subsistence, equipment, waste management, travel; approaches to the sharing economy).

Networking

- industry initiatives (documents focusing on policy; the use of best practice guides to normalize sustainability);
- workgroups (participation at seminars and workshops organized by industry organizations – British Academy for Film and Television Arts, Ecoprod);
- consultancy organizations (the role of commercial ventures devoted to developing sustainability policies)
- organic linking (networks between individuals and collaboration from a grassroots perspective)

These categories are only a sample of a much wider body of concerns, and while these areas overlap considerably in practice, the brief outline above provides an overview of the key areas on which this book will focus.

Rationalizing the need for environmental management of the media

While academia has slowly embraced the necessity of addressing the footprint of the media, this trajectory has taken a somewhat different but similarly complex path in the media industry. Traditionally, media companies have found it very difficult to conceptualize appropriate measurements to understand their impact on the environment, and as a consequence, the perception of the media industry as a sector with a comparatively insignificant footprint has persisted. This assertion is largely based on comparison with heavy extractive industries, as their scale of harmful emissions visibly exceeds those from the media sector. Some estimate that the scale of emissions comparing media with extractive sectors comes to an equivalence of 1:143 (Media CSR Forum 2013). While organizations like the US-based Environmental Media Association (EMA) have lobbied in favour of sustainability issues since the late 1980s, many parts of the industry have struggled to establish the validity of the issue for their stakeholders, and thus establish concrete strategies to both account for and act on emissions.

The work of the Responsible Media Forum (also known as the Media CSR Forum), a UK-based organization providing consultancy services for different areas of the media industry, provides a good example of the challenges facing the establishment of environmental initiatives. The venture was established in 2001 as, prompted by the charity organization Business in the Communities, certain media companies took part in an exercise to assess the sector's understanding of environmental responsibilities (KPMG 2005). The results were disappointing as the sector did not have any real strategies in place to evaluate the impact they had on the environment or the kinds of emissions they generated. Business in the Communities commissioned a materiality exercise from KPMG that consulted individuals and organizations in and outside of the media to evaluate where the material expenditure of the sector lies and what kinds of emissions were generated by different media companies. These are often very different from an oil or construction company, who have clear material expenditures and a very visible carbon footprint, and consequently, must work to manage their 'green' reputations. Ironically, in terms of reporting on emissions, the media companies fared poorly against well-known polluters like BP and Shell.

The parent company of the Responsible Media Forum, Carnstone, has done an exercise on the carbon impact of the UK's ten largest media and extractive companies (Media CSR Forum 2013). The results show that the comparative statistics between the scale of the emissions are in a completely different league. In terms of employee numbers and financial investment, the media and the extractive industries tend to follow similar patterns of scale, but emissions show a very different comparative picture. According to Carnstone, the impact of all the media companies in the world would only total a small proportion of the carbon impact of the extractive sector. For Carnstone, this would be reversed if we focused on the impact of content – that is, what they call the brainprint (ibid.). This juxtaposition is an intriguing reflection of the ways academia has responded to environmental concerns to do with the media – content dominates, with only relatively meagre interest in the footprint. Thus, it is not surprising that most studies from industry and academia have focused on the ability and responsibilities of the media to communicate about the environment.

As we will see below, this indifference is not only relegated to consulting companies but also the whole industry chain of the media, from regulation down to individual players. Simply put, these are not perceived as 'material issues that are financially significant or affect a key financial indicator of the industry' (ibid.: 5). Nor are they strategic in that they significantly affect the ability of the company to deliver its strategy. From the perspective of the Media CSR Forum, they are only operational issues that 'matter for other reasons – internal, reputational, efficiency' (ibid.: 5). For them, these areas do not present a significant threat to the company in the way that the oil industry addresses them. For Shell or BP, the issue of sustainability is a huge concern as regulatory regimes and increased investment in green technology are altering the fundamental production and financial foundations of the industry. For media, the emphasis on green technologies and sustainable protocols does not meet this level of complex existential urgency.

In part, the aim of this book is to evaluate and counteract this argument. I will lay out the rationale for addressing the environmental impact of the sector, including establishing clear policies on footprinting, but two arguments shall suffice for now. The first of these concerns the extent of emissions from commonplace activities in the sector, especially ones that would seemingly be exempt from large emissions scales. For example, the British Academy of

Film and Television Arts (BAFTA) estimates that a single hour of television drama produces 13.6 tonnes of CO_2. This is the equivalent of approximately four times the size of an individual's annual CO_2 footprint. Second, media texts are often commended for their awareness generation – even by organizations like the Forum – yet can they justifiably communicate these messages if they do not practise what they outwardly preach? The material and the ethical questions these two areas raise comprise two particular argumentative strands that will be explained systematically throughout this book.

Complicating these arguments is the omnipresent economic logic of industry discourse. It pervades in particular the concept of sustainable development, characterized by the Brundtland Commission as 'development that meets the needs of the present without compromising the ability of future generations to meet their own needs' (1987: 2). Ever since its popularization in the wake of the Commission's focus on the term, it has been used as an aspirational model by academia and industry. But this is a concept enmeshed in substantial conceptual blurriness. An initial complication comes from its application to cover economic, societal and environmental development, where its use in general is often interchangeable on all three areas of sustainability. Another concern comes from deciphering where the balance between the emphasis on sustainability and development falls. The latter comes with a heavy emphasis on economic baggage where business and efficiency concerns can easily overtake more environmental priorities. As we will see, the logic of economic development pervades discourse in the media industry and, consequently, indicates a certain dilution of their notions of and practices for incorporating sustainability. For example, in response to a question over why SVT, the Swedish national public service television company, has engaged sustainability strategies, the reply from their communications director is surprisingly blunt. The adoption has been 'done primarily for economic reasons' (Ahlstrand, pers. comm., 9 April 2017).

While other companies have more moderate views on the topic, the financial challenges and opportunities that environmental sustainability can provide comprise an area that cannot be ignored. The media is a business, of course, and it would be counterproductive to treat it as anything but an economic enterprise. Yet, consigning its focus on environmental sustainability to be merely operational or reputational – to use the designations of the Forum – emphasizes the sense that sustainability is an afterthought, a non-essential area of operational concern with no impact on the bottom line. Or to put it in other words, if sustainability is consistently marginalized by the economic and material necessities of industry development, how feasible is it to merge sustainability and development in an industry where its key performance indicators (KPI) emphasize both efficiency and expediency as optimal values? Can sustainability and the media business co-exist productively?

In extreme variations of this argument, any notion of sustainable development is akin to an oxymoron. For example, from the perspective of ecomarxist commentator Joel Kovel, any entanglement with capital is detrimental to environmental argumentation. He adopts an ecosocialist perspective that argues that:

Capital in its essence is not directly part of nature at all. It is a kind of idea in the mind of a natural creature which takes the external form of money and causes that creature to seek more of what capital signifies. It is this seeking through economy and society that degrades nature. Capital becomes both a kind of intoxicating god, and also a force field polarizing our relation to nature in such a way that spells disaster.

(Kovel 2002: 3)

Kovel's perspective may seem somewhat over-the-top in its view on capitalist ideology, but it does raise pertinent questions about the ways nature is 'severed from its ecosystemic existence and ... subjugated to human agency. Thus, the ceaseless rendering of the environment into commodities, with its monetization and exchange, breaks down the specificity and intricacy of ecosystems' (ibid.: 40). Kovel's arguments share much with ecocritical work exploring film and television content, especially on how they use ecological themes to address exploitation of peoples and the environment. Critics focus on areas such as resource use, land ethics, food politics, and the commodification of lifestyles normalized as conventional behaviour through narratives of films and other media. The media's integration with the environment is here considered in positive terms as it can be used to address anthropocentric approaches to the environment.

But from the perspective of media management, capitalism and commodification are, of course, part of the standard operating procedures of the industry. A pragmatic view of integrating these procedures with a more ecocritical perspective balances between a realistic understanding of the media as a business and the types of challenges that environmental sustainability poses for its practical implementation to industry management. The point here is not to apologize for the, perhaps, inevitably flawed take of the industry on sustainable development but to provide an analytical position with a practical approach. Ecosocialist rhetoric, for example, would need to be adapted and integrated by considering the KPIs of organizational management and the political economy of the industry if we are to provide a fully critical and balanced evaluation of how environmental regulation and production operate in practice, as well as the challenges that environmental sustainability faces. This does not mean that we can just ignore the problems the sector is causing, but instead that we must focus on addressing the strategies the sector has in place and in development to best evaluate and understand the role the environment holds within it.

Bibliography

Albarran, Alan (ed.) 2005. *Handbook of Media Management and Economics*, New York: Routledge.
Anderson, Alison. 2014. *Media, Environment and the Network Society*. Basingstoke: Palgrave.

Anderson, Alison. 1997. *Media, Culture and the Environment*. New York: Routledge.

Appadurai, Arjun. 1996. *Modernity at Large: Cultural Dimensions of Globalization*. Minneapolis, MN: University of Minnesota Press.

Bennett, Jane. 2010. *Vibrant Matter: A Political Ecology of Things*. Durham, NC: Duke University Press.

Bozak, Nadia. 2012. *The Cinematic Footprint*. Piscataway, NJ: Rutgers University Press.

Brereton, Pat. 2005. *Hollywood Utopia: Ecology in Contemporary American Cinema*. Bristol: Intellect.

Brundtland Commission. 1987. *Our Common Future*. Oxford: Oxford University Press.

Cox, Robert and Pezzullo, Phaedra. 2015. *Environmental Communications and the Public Sphere*. Thousand Oaks, CA: Sage.

Cubitt, Sean. 2005. *Ecomedia*, Amsterdam: Rodopi.

Cubitt, Sean. 2017. *Finite Media: Environmental Implications of Digital Technologies*. Durham, NC: Duke University Press.

Fuchs, Christian 2008. *Internet and Society: Social Theory in the Information Age*. New York: Routledge.

Fuchs, Christian 2014. *Digital Labour and Marx*. New York: Routledge.

Gabrys, Jennifer. 2011. *Digital Rubbish: A Natural History of Electronics*. Ann Arbor, MI: University of Michigan Press.

Good, Jennifer. 2014. *Television and the Earth: Not a Love Story*. Halifax: Fernwood.

Greenpeace. 2017. *Clicking Clean: Who Is Winning the Race to Build a Green Internet?* Available at: https://secured-static.greenpeace.org/austria/Global/austria/dokumente/Clicking%20Clean%202017.pdf (accessed 11 November 2017).

Gustafsson, Tommy and Kaapa, Pietari. 2013. *Transnational Ecocinema: Film Culture in the Age of Ecological Depravation*. Bristol: Intellect.

Hansen, Anders. 2011. *Environment, Media and Communication*. New York: Routledge.

Hibberd, M. and Nguyen, A. 2013. 'Climate change communications and young people in the United Kingdom: a reception study', *International Journal of Media and Cultural Politics*, 9(1): 27–46.

Hodgson, Christopher. 2015. 'Can the digital revolution be environmentally sustainable?', *The Guardian*, 13 November 2015.

Ingram, David. 2000. *Green Screen: Environmentalism and Hollywood Cinema*. Exeter: University of Exeter Press.

Ivakhiv, A. J. 2007. 'Green film criticism and its futures', *Interdisciplinary Studies in Literature and Environment*, 15(2), 1–28.

Ivakhiv, A. J. 2012. 'Teaching ecocriticism and cinema', in Garrard, G. (ed.) *Teaching Ecocriticism and Green Cultural Studies*. Basingstoke: Palgrave Macmillan.

Kääpä, Pietari. 2011. 'Ecocinema audiences', a special issue of *Interactions: Studies in Communications and Culture*, 4(2).

Kääpä, Pietari. 2014. *Ecology and Contemporary Nordic Cinema*. New York: Bloomsbury.

KPMG. 2005. *The Media CSR Forum*. London: KPMG.

Kovel, Joel. 2002. *The Enemy of Nature: The End of Capitalism or the End of the World?* Black Point: Fernwood Publishing.

Lu, Sheldon and Mi, Jiayan (eds) 2009. *Chinese Ecocinema: In the Age of Environmental Challenge*. Hong Kong: Hong Kong University Press.

Maxwell, Richard and Miller Toby. 2012. *Greening the Media*. Oxford: Oxford University Press.

McLuhan, Marshall. 1964. *Understanding Media: The Extensions of Man.* New York: McGraw-Hill.

Media CSR Forum. 2013. *Mirrors or Movers.* London: Media CSR Forum Secretariat.

Parham, John. 2015. *Green Media and Popular Culture.* Basingstoke: Palgrave.

Parikka, Jussi. 2015. *A Geology of Media.* Minneapolis, MN: University of Minnesota Press.

Parks, Lisa. 2007. 'Falling apart: electronics salvaging and the global media economy', in Acland, Charles (ed.) *Residual Media.* Minneapolis, MN: University of Minnesota Press, pp. 32–47.

Rust, Stephen, Monani, Salma and Cubitt, Sean (eds) 2014a. *Ecocinema: Theory and Practice.* New York: Routledge.

Rust, Stephen, Monani, Salma and Cubitt, Sean (eds) 2014b. *Ecomedia: Key Issues New.* York: Routledge.

Starosielski, Nicole and Parks, Lisa. 2015. *Signal Traffic: Critical Studies of Media Infrastructures.* Chicago: University of Illinois Press.

Turkel, Sherry. 2008. 'Always-on/Always-on-you: the tethered self', in Katz, James (ed.) *Handbook of Mobile Communication Studies.* Cambridge, MA: The MIT Press.

Walker, Janet and Starosielski, Nicole (eds) 2016. *Sustainable Media.* New York: Routledge.

Willoquet-Maricondi, Paula (ed.) 2010. *Framing the World: Explorations in Ecocriticism and Film.* Charlottesville, VA: University of Virginia Press.

2 The network

Introduction

Any analysis of environmental sustainability and media management must take into account a multitude of agents and power relations that form a network of influences and complex dynamics. The process of establishing policy and best practice is a negotiation, determined by a range of political and economic interests as well as the material realities that underpin these negotiations. Simply focusing on the political economy of media management would only permit us to focus on human actors as they negotiate network dynamics. Environmental management is a lot more complex in orientation and must take into account the grounding realities of material concerns that contribute to the environmental impact of media production and consumption. Clearly, a more complex framework of analysis is required.

Félix Guattari's 'three ecologies' (2000) provides a productive starting point for such an analysis of environmental management of the media. Guattari's work on what he calls the three ecologies is concerned with the ways in which environmentalist thinking is often impeded in its incorporation into human activity and social organization. To evaluate what hinders this incorporation, the analyst must focus on three areas where the environment and the human intersect. These consist of the interacting and the interdependent spheres of the mental, social and the environmental, or the cognitive, the political and the environmental. For Guattari, these three areas consist of distinct ecologies, of ways of comprehending relationships with the external world. To succeed in facilitating transformations in perception and action, human awareness of their own role in the environment, the political organization of society and environmental awareness need to come together holistically to generate a shift in perception. Guattari argues that adopting environmentalism is not enough as 'ecology must stop being associated with a small nature-loving community. Ecology in my sense questions the whole of subjectivity and capitalist power formations' (ibid.: 35).

For such a transformation in ontological perceptions to take place, environmental awareness must be generated on all three levels and recognize the specific conditions of each ecology as well as the rules that govern them. To explain, human subjectivity does not automatically identify with, or agree to, the most productive social welfare or sustainable behaviour. Guattari suggests that this is often the opposite as political concerns override mental awareness of issues such as sustainability, which, in his work, is a particular result of capitalist hegemony on the constitution of normative mental and social ecologies. Instead, it is the 'job' of an ecophilosophical argument to bind the three ecologies together in ways that lead to productive outcomes while accounting for the diverse requirements embedded in each perspective. In practice, this would take place by ensuring that these connections use the particular qualities of each ecology. For cognitive challenges, the focus would be on new ways of thinking and perception. For society, this involves a radical politics that challenges the patterns of normalization that regulate, for example, the adoption of sustainability initiatives in the industry. And for an environmental perspective, the emphasis is on highlighting connections that enable a more productive perspective on sustainability to materialize. For this sort of correlation to take place efficiently, I will adopt the concept of the network as a productive tool of analysis as it opens up the possibility of addressing the multitude of influences and power relations necessary to establish environmental sustainability as a viable media strategy.

Networks

Guattari's emphasis on correlating different ontological realities to form a holistic view of the-human-in-the-environment is the starting point for understanding the media in its ecosystemic context. John B. Thompson's perspective of the production of media as a contextual process allows us to position Guattari's approach more comprehensively with the technological and social processes of the media industry:

> Mediated communication is always a contextualised social phenomenon: it is always contextualised in social contexts which are structured in various ways and which have a structuring impact on the communication that occurs. Since mediated communication is generally fixed in a material substratum of some kind, it is easy to focus on the symbolic content of media messages and to ignore the complex array of social conditions which underlie the production and circulation of these messages.
>
> (Thompson 1995: 11)

Thompson's arguments are mostly concerned with the complexities of increasing connectivity and globalization, but his emphasis on material conditions also previews many of the arguments about contemporary network theory, as well as the emphasis of ecomedia studies on material processes. In describing the ways in which social contexts condition the behaviours of individuals within them, he draws on Pierre Bourdieu's fields of interaction.

Here, individuals have to pursue their objectives and opportunities according to the limits of their circumstances. Depending on the position they hold in these structures and the resources they have available, they have varying levels of power to influence the construction of the particular network in which they operate. While, for Thompson, interactivity and connectivity are ways to make sense of how individuals operate in certain cultural and social circumstances, for us, they provide the basis for understanding environmental media management as a network of complex power relations. Instead of understanding it as a strictly hierarchical or linear form of organizational management, my approach to the concept of the network is concerned both with organizational management as well as the blurring of the human-nature dichotomy explicit in new materialist studies.

The first step is to chart the ways in which different parts of a media organization come together in the wider media environment. The advantage in perceiving this as a network comes from moving beyond a traditional hierarchical model in which organizational management is often premised to network-based analysis of the relationalities of power and influence in organizations. Second, to facilitate an understanding of the environmental complexities of media practice, our comprehension of the network will have to be much more complicated than simply focusing on anthropocentric managerial aspects. In such a perspective, regulators, managers, producers, accountants, technicians are just one part of this network. Here, theoretical work from new materialist initiatives allows us to construct an enhanced conception of the network that is as non-anthropocentric in its constitution as possible. Such a network aspiring to environmentally sustainable media production is best described as a process relational system, meaning that its constitution operates as a complex set of material relations where its material base, consisting of a variety of metals, chemicals, minerals, energy sources and other substances, plays a role alongside human-led managerial decisions.

Allowing matter to have 'agentic' potential in network processes moves us closer to what Guattari was aiming for with his conception of the three ecologies, that is, a more comprehensive understanding of the-human-in-the-ecosystem. An initial starting point for this sort of analysis comes from the work of McLuhan (1964) and Neil Postman (2005), especially, on interconnections between media objects and technology that gestures to understanding technology as an extension of the human. The integration of the human with an external world

> is meant to point to the lack of a beginning and end point in a system that is explained not only by interactions between humans and technologies, but rather in addition by a continuous, co-creating exchange. To put this in the terms of media cultural studies, we might say that humans and technologies are both simultaneously actors and acted upon, even if often temporarily, and occasionally with unpredictable results.
>
> (Lundby 2009: 95)

While the focus here is very much on technology, the approach lays the foundations for a perspective that understands the materiality of the media as a complex layering of processes and material connections, of assemblages of natural elements that both comprise of and leave material traces. Seen through the lens of media archaeology, a field of study uncovering the 'weird materialities' (Parikka 2012) that enable the construction of new media devices as well as their integration into the social fabric, McLuhan's and Postman's emphasis on the reciprocal qualities of processes between technology and humanity can be seen to gesture to the idea that any analysis of the media 'needs to insist both on the material nature of its enterprise – that media are always articulated in material, also in non-narrative frameworks' (ibid.). As Parikka suggests in *A Geology of Media* (2015), the materials of which media devices and means of production and dissemination consist have histories of their own. The metals, minerals, chemicals, toxins, and so on exist autonomously before, during and after their incorporation into the 'media', and need to be perceived as such, as agentic materialities with a considerable role in the media production network.

This approach leads us to consider the relationship between human agency and other materialities in more complex terms than foregrounding anthropocentric concerns – while acknowledging the central organizing role these concerns hold. After all, the mental and the social are key ingredients in the holistic ecologicalism that Guattari endorses. What I have in mind is combining media management with the work of Matthew Fuller (2005) and Jussi Parikka, among others, who explore 'the massive and dynamic interrelation of processes and objects, beings and things, patterns and matter' (ibid.: 2). They practise a form of media ecology that aims to provide an understanding of the totality of interconnections that facilitates the construction and dissemination of the media, human and non-human alike. Erasing distinctions and power differences between the different components of a network is the goal here as this form of media ecology aims for 'an address of the materiality and immateriality of media objects, devices, and systems in terms of their form as both pattern and presence … Nonlinear, self-organizational, and transpositional systems behaviour combine autopoietically at the intersection of media collisions' (Slayton 2005: 1). The emphasis in this line of inquiry is on complex relations of agency, power, organization and intent in these networks of people, technology, resources, managerial directives, as well as matter that hosts agentic power of its own. This approach aims to place the media firmly in its ecosystemic context.

The central significance of material agency in understanding how the media production network operates can best be explained by focusing on examples from media production and distribution management. In discussing the constitution of the systems facilitating telecommunications, Rahul Mukherjee suggests that 'these mediating infrastructures are produced not only through the agencies of human stakeholders, but also via nonhuman actors and dynamic materialities' (2016: 99). Here, Mukherjee suggests that material aspects of the communications infrastructure, including cell tower antennas and the radiation they emit, need to be considered as part of the political ecology of the society for which they facilitate communications. Human-led construction and organization of the communications network are only a part of the picture as this does not cover what I call the material base of the media infrastructure. Similarly, all the resources that go into film production share this form of material agency. These can be the petrol that fuels the vehicles, the electricity that powers the cameras, the food and water that sustain the crew or the wood and metals that comprise the sets. The procurement of these materials is achieved according to the practicalities and economics of a film budget and are thus coordinated according to anthropocentric logic.

Yet, as Bennett suggests in *Vibrant Matter* (2009), material objects have to be considered as active participants in any act of becoming, instead of being seen only as static and passive substances. The material has a 'say' in how it is incorporated into the production as their agentic qualities influence choices made on set. If we take LED lights as an example, we can see how incorporation of LEDs as standard practice changes operational procedures. The advantages from this change of practice come from LEDs being much more energy-efficient than standard lights. Furthermore, they take up much less space in set construction and organization. Yet, changing all the existing light bulbs to LEDs would be a costly venture and would impact on the economic bottom line of a film or television. The material qualities of LEDs thus pose a substantial challenge to the standard operating procedure of film and television production. In response to these types of material challenges, BAFTA runs a workshop on carbon literacy where they, for example, discuss the adoption of LEDs as an industry standard. The arguments for and against adopting LEDs as standard practice not only highlight production practice but also reflect on the aesthetic qualities of the final product – the content – as the quality of light from LEDs tends to be less pronounced than from ordinary bulbs. Yet, most productions are graded in post-production, and thus any of the negative impacts that the more resource-effective LEDs may have on the quality of the product can be reduced. Thus, the adoption of LEDs for a production are – yes – mandated by the decisions of producers and craft personnel. But they are also predicated on the particular material qualities of the LED technology, which influences both practice and the financing of a production.

Following the new materialist work of Karen Barad, Diana Coole and Samantha Frost suggest that 'materiality is always something more than "mere" matter: an excess, force, vitality, relationality, or difference that renders matter active, self-creative, productive, unpredictable' (2010: 8). Parikka calls these elements that play a key role in the production and distribution network but are not explicitly part of the text, 'nonmediatic materialities' (2013: 70). These are areas that exist externally to the media as consumers see it, but which constitute a fundamental part of its ecosystemic presence. From a network perspective, these are all essential components that dictate in part how the network operates as a system and how it facilitates new connections. Taking the case of LED as a particular example of agentic matter highlights the ways we must afford objects used in the creation of texts a level of agency. This exceeds the notion of production materials as only passive objects that accumulate relevance through their application by the work of the human mind. Even as the construction and application of LEDs are dictated by human agency, the ways they influence production decisions emerge from their materiality.

Yet, while such an approach is essential for a more comprehensive understanding of the complex relations of production decisions, we must also remember Christian Fuchs' point that for a practical understanding of environmental problems, we need to view them as social problems. Thus, a new materialist perspective on media production is a process of 'a doing, a congealing of agency, a stabilizing and destabilizing process of iterative intra-activity' (Barad 2007: 151). I take this to mean a reciprocal sense of dialogue between human and non-human agency. To explain, in the case of LEDs, this entanglement of agency can be witnessed in the ways the material properties of LEDs feed into production management, which requires consideration not only of material and technological advances but also of transformations in labour practice, and not only of environmental benefits, but also of initial expenses in adopting new technology as well as the eventual cost savings from adopting them.

Thus, a focus on environmental management of the media would need to take on board Serenella Iovino and Serpil Oppermann's definition of material ecocriticism that focuses on 'matter's narrative power of creating configurations of meaning and substances that enter into human lives in a field of coemerging interactions' (2012: 79–80). Such a form of ecocriticism must be premised on a reciprocal view of material processes and managerial strategies. I am most interested in the ways materialities are translated into policy and practice in the forms of texts that can be read, I argue, to understand the process relational networks that comprise the practice of environmental management of the media. These texts are comprised of industry documents on areas like regulation and training and consist of strategy and policy publications, as well as a whole range of industry manuals and promotional material. While these are very much in the anthropocentric realm, the agentic power of materialities can, arguably, be traced through the mobilization of specific framings and rhetoric. While this, of course, can be seen to limit non-human matter's agentic capacities, it also reflects industrial realities where material elements are always incorporated through a lens that emphasizes their relevance for the key performance indicators (KPIs) of the industry. Thus, they are often secondary to the material and strategic realities of an industry focused on producing texts in ways that conform to budgetary and regulatory requirements. A pragmatic and dialogic perspective would need to incorporate theoretical frameworks such as new materialism with an understanding of industrial realities in ways that consider materiality and management as part of the same realm of enunciation.

At the same time, there is no point in pretending that the industrial realities of media production comprise a level playing field, at least in terms of 'translating' material agency into policy. When exploring the agentic presence of materialities in production networks, we often have to go through translated recontextualizations, such as documents and descriptions of the ways that productions, for example, have integrated material concerns into operations. The process of translating materialities into protocols and practices results in the enforcement of anthropocentric principles on these materialities. Often, this leads to a situation where, as Maxwell and Miller point out, 'in the terminology of ecological ethics, we might say that a dominant anthropocentric aspiration insists on managing environmental risks in ways that prioritise established human interests' (2017: 175). The case of LEDs, for example, can illuminate this as the benefits of energy savings are converted into economic and managerial cost evaluations that can impede any appropriate adoption of LEDs. This conversion of energy flows to managerial practice codifies materiality with conceptions such as temporality and efficiency to situate them according to the logic of a business company. The process of translation can be seen to derive agency from materialities and redistributing it to the human components of the network, an act which comprises a form of 'reterritorialization'. The coding of materialities in the language of economics and management provides agentic territory for human participants, shifting the power to coordinate the network away from its material base. These acts of reterritorialization operate as discursive devices to counter the challenges more nuanced perspectives on sustainability would offer. As we will see, it is these acts of anthropocentric reterritorialization that dominate much of the industry rhetoric.

Yet, as geographers Bruce Braun and Sarah Whatmore remind us, objects 'cannot be reduced to things on which decisions are made in the political realm because they are part and parcel of that realm from the outset' (2010: xxii). Similarly, Janet Walker and Nicole Starosielski suggest that 'an ecological media materialism always addresses more than just the mere physicality of media infrastructure, technologies and objects' (Walker and Starosielski 2017: 13). Both perspectives emphasize the complexity of evaluating the fault lines between the human and the non-human. Continuing this line of inquest in outlining a material form of ecocriticism, Serenela Iovino and Serpil Oppermann (2012) suggest that the dynamics between the agentic qualities of material and the anthropomorphic tendencies of narrative actions need to be seen as a form of dialogue. The consequence of this would be a polyphonic act revealing both similarities and relations between the ecocentric and the anthropocentric. This would be the ideal state of analysis, even for a study that focuses on policy documents, but as suggested above, industrial realities do not fit easily with such projections of equality between the human and the non-human. Certainly, all acts of human narration have their roots in material conditions, but there is often a huge disconnection between these roots and the complex objectives of the industry. It would be difficult to rationalize placing these two at an even level in an industry that is, in many ways, disassociated from adopting environmental sustainability as standard practice.

The disconnection of anthropocentric management and ecocentric materiality is a more pervasive problem than simply an industry looking after meeting its KPIs. Cubitt (2017) suggests that we have been trained to perceive the world as an ontology of things instead of a complex connective materiality. This testifies to the pervasiveness with which reterritorializing discourses reverberate through the three ecologies of human activity. The pervasive dominance of anthropocentric logic and the obstacles it causes for trying to reposition our three ecologies along more environmentalist lines would require an ontological reorientation in our thinking. To achieve this reorientation towards a more complex understanding of material networks, we would need to find a balance between agentic materiality and human influence. The problem is one of discourse, considered here in its Foucauldian sense as a means to organize power structures. The act of reterritorialization, for example, is a particular discursive strategy that wrests control of the network for anthropocentric interests. To understand best how environmental sustainability is integrated into media practice, we have to evaluate how discursive framings operate in the management and production of media.

Actor-network theory and the material network

The analysis of these material networks cannot be coordinated through conventional approaches to the political economy of the media. A more conducive approach comes from political ecology, an approach that investigates not only the influence of political and economic forces on media infrastructures but emphasizes their environmental and material roots. In order to achieve this methodologically, I adapt a range of techniques from actor-network theory (ANT). This is an approach that emerged from the work of John Law and Bruno Latour to explain how networks generate meanings and actions. Analysis of actor networks focuses on power relations between different components that constitute the network to facilitate a better understanding of how such a network comes to be and what maintains its constitution. A key area distinguishing ANT from other systems-based approaches is that there is more to the network than a conventional human-led organization of participating actants. In addition to societal and organizational players – the anthropocentric world – the role of objects and material processes receives equal attention as part of the network dynamics. Most of the key ideas discussed above – process relationality, material becomings – feature here as ANT has been used to 'explain social order and sociality within actor-networks, unearth the deeper understanding and complexity of agency, address questions about competing values among network actors, and highlight the need to establish priorities in approaching' (Luppicini 2014: 41). ANT is particularly appropriate for this study of environmental media networks as it can be used to emphasize the complex material base that informs and shapes managerial decision-making practices. Consequently, ANT is used here for exploring the environmental management of the media by focusing on the interactions of regulation, management, practice, communications and the material base.

The connections between the different parts of the network provide some of the key analytical foci for our study. In an actor network, each component, or actant, is connected with a range of other similar actants – Law (2007) even suggests that each actor is a network unto themselves. This makes the mapping of a network an infinitely complex undertaking, even as it also underlines the need to focus on all aspects of the network. The points of contact can be considered 'nodes' that connect exponentially to other agentic elements to produce processes that generate the meanings the network emits externally. Each node can have as many dimensions as they have connections, as these connections allow them to contribute to the network in unexpected ways. To take an example, in an environmental media management network, a node would be a central organization like BAFTA. This organization would generate its own network based on its sustainability principles and connect to a range of production companies, non-governmental organizations (NGOs), regulatory institutes, and suppliers of goods and services. In addition, it would be part of not only a wider network for environmental media – connected at an institutional level with other similar organizations, such as the Arts Council England – but also play a part in a much more comprehensive network on environmental incentives in the UK, comprised of representatives from different sectors and from the regulatory establishment.

The key networking tool for BAFTA to use would be a particular strategy or even an incentive that acts as a significant locus for the industry. In actor-network terminology, such elements would comprise an obligatory passage point (OPP), a term used by Law to describe elements through which all aspects of the network will have to pass, a sort of a funnel used to process elements into their optimal positions in the network. In the case of BAFTA, this would comprise its bespoke carbon calculator Albert, which it has designed in collaboration with the NGO Julie's Bicycle. Albert is free to use for domestic productions and allows companies to calculate their emissions through CO_2 equivalents. Television production companies operating in the UK media environment would often need to go to BAFTA to obtain finance for their productions and would thus be required to use Albert as part of their accounting. The Albert OPP thus connects a multitude of participants in the UK media environment from the production companies that use it to report on their emissions to companies in the supply chain whose services are measured with its tools. Albert can also connect the industry with institutional organizations like the BBC (who also use it), NGOs such as Julie's Bicycle, who have participated in designing it, to the Department of Sports, Culture and Media that sets policy for the sector. It is thus involved with a range of stakeholders as well as the material base – the emissions the media generates – that comprises the data for the reports it is used to generate, and which influences a company's understanding of its own footprint – in addition to its role in the wider UK environmental media network.

Another key consideration in adopting an actor-network approach emerges from the potential to see it as a discursive structure comprised of dynamic autopoietic relations. Paul Bourgine and John Stewart define an autopoietic network as 'a network of processes that produces the components that reproduce the network, and that also regulates the boundary conditions necessary for its ongoing existence as a network' (2004: 327). BAFTA operates as this kind of self-regulating network that produces the materials for the network – for example, its OPPs – and sets its boundaries. Establishing these conditions for a media management network would be a strategic task for an organization like BAFTA and require the formulation of company policy to ensure all its stakeholders and external collaborators operate with the organization's environmental sustainability goals in mind. The formulation of tasks and boundaries to construct the network can be observed by the variety of discursive elements of policy that outline the processes that shape, constrain, formulate, and limit the existence of the network. In practice, this means all the various regulations, codes, procedures, standards and communications that flow internally and externally between different nodes of the network.

The focus on policy necessitates exploring translational areas such as discourse and rhetoric – processes that translate materialities into networking codes – to understand the entanglement of both material and human-led agency. To study these areas, I focus on organizational documentation and practice to interrogate the extent to which the processes of de- and reterritorialization emerge in the rhetoric of media organizations. Key parts of the discussion will focus on discourse analysis of regulatory statements and interviews with regulators and media professionals. These areas – the regulatory environment and specific strategic statements from various stakeholders – comprise the translational reterritorializing processes that will be evaluated to ascertain the balance of the network. Law reminds us that all discursive patterns are always the outcome of a number of agents intertwining as part of a network:

> The actor network theory is distinctive because it insists that networks are materially heterogeneous and argues that society and organization would not exist if they were simply social. Agents, texts, devices, architectures are all generated, form part of, and are essential to, the networks of the social. And in the first instance, all should be analyzed in the same terms. Accordingly, in this view, the task of sociology is to

characterize the ways in which materials join together to generate themselves and reproduce institutional and organizational patterns in the networks of the social.

(Law 1992: 379)

This reproduction of patterns is predominantly conducted by what I call 'ordering discourses' a concept that draws on Foucault's *The Order of Discourse* (1981) to identify how discourses can structure social and political relations based on the value systems embedded as part of their utterance. Discourse acts as a materially inflected means of communication where discursive acts organize networks according to specific orders. Operating as a tool of power, we need to consider its role as a means to wield control over the network. ANT is particularly useful here as it facilitates analysis of the ways in which discourse is inevitably tied up with the material conditions of its utterance. Thus, we need to position discourse in its contexts to decipher how discursive acts order materiality and how these materialities can shape or resist the manipulations of discourse. By emphasizing the agentic role of these materialities, this allows us to evaluate the processes of reterritorialization in these discursive practices.

Discourse analysis is an appropriate means to uncover the rhetorical tools and coded linguistic structures – such as those belonging explicitly to business or management – used by human agents to control the network. By focusing on discursive structures and texts as essential parts of the network, two types of ordering discourses can be identified. The first is an ecocentric one that comprehends human materiality as only one part of the network. Ecocentric perspectives approach humanity as part of the diversity of the ecosystem, on a par with all the other organisms and processes that comprise its complex structures. The concept of 'the more than human' world indicates the basic tenet of an ecocentric worldview, where humanity is just one component of a much more complex network of intersecting, inter-reliant existence. While such rhetoric is inevitably anthropogenic, these arguments are conducted in a way that balances the environmental and the anthropocentric to displace the human from its position as the locus of all meaning.

The second type of ordering discourse concerns anthropocentric rhetoric that prioritizes societal or economic benefits over other parts of the ecosystem (Räikkönen 2011). Such perspectives consider the welfare of humankind as the key principle for planetary development. In many ways, the logic and argumentation of sustainable development are an embodiment of this rhetoric, which is a specifically significant concern for us as it functions as a key ordering discourse to integrate sustainability into media production networks. These are the types of discourses that prioritize competitiveness and costs and argue for economic sense in developing or adopting green standards. As a pragmatic necessity, much of the rhetoric by both activist organizations campaigning for greening the media and policy documents from the industry base their arguments on the financial savings that can be developed through sustainability measures. This is an important tactic to adopt for an industry that is still inherently a business and where any new expenditures – both in terms of money and time – are a potential problem for delivering on its KPIs, especially on producing content. Anthropocentric arguments frame sustainability measures as a logical step for the industry, where it is essentially pitched to producers as a cost-saving mechanism or a means to generate positive publicity. Seen from an ecocritical perspective, this appropriation of environmental terminology and rhetoric 'colonizes' the materiality of the network and provides an act of reterritorialization against the environmentalist inclinations of sustainability.

Actor-network theory provides a productive approach to evaluating the eco- and the anthropocentric connotations of the management discourses used by the media industry. While the human inevitably speaks for the network in ways that could potentially limit the agentic materiality of objects and matter, the focus of analysis must be on the extent and means of material to dictate the vernacular. In reality, this is often limited by industrial realities where the environment only occupies an operational importance. Thus, it is necessary to adopt an ecocritical perspective on the anthropocentric rhetoric that dominates the ordering discourses of these management networks. This is especially the case as managerial discourses tend to use vocabulary from ecocentric rhetoric. Such vocabulary provides both credibility for managerial strategies while also arguably serving as greenwashing rhetoric to mask their roots in explicitly anthropocentric areas of concern.

In focusing on discourse analysis, I draw on Mukherjee's (2016) emphasis on a radical entanglement of materiality and discourse, which argues that we must explore not only the infrastructure but also the act of mediation about that infrastructure. Such a perspective turns our attention to policy documents, which act as the means of ordering a network to meet the requirements of its key organizations. These mediations of infrastructure can provide a sociomaterialist understanding of that infrastructure in all its relationalities as well as uncover its power relations. Policy documents can be approached from an ANT perspective through both qualitative and quantitative means. Here, the connections of the network can be analyzed by both the substance – the rhetoric – that forms them as well as their volume – the number of times certain translational conventions are used. Thus, my role is to observe the discursive means that connect the nodes and explain how power relations between the agentic components of the network are constructed. The focus is on analysis of regulatory directives and managerial codes to reveal the complex reterritorializing mechanisms of the autopoietic practice of the network. Exploring management and regulation enables us to focus on the ways the boundaries of the network are organized, and how different power relations of the network influence the processes that take place in it.

Case study: the Environmental Media Association (EMA)

Perhaps the best way to explain the practicalities of adopting aspects of actor-network theory to study environmental media management is to explore a case study of the Environmental Media Association (EMA). The first organization to focus exclusively on the environmental role of the sector, the EMA was established in Los Angeles in 1989 to address the environmental impact of the media sector – here defined as television, music and film. Impact in this context was based on evaluating the effectiveness of the sector in generating awareness of environmental issues. The activities of the EMA were based on communicating messages about the environment as 'the entertainment industry can reach millions of people with a message of concern about our environment and examples of concrete action individuals can take to make a difference' (EMA 2017a). As part of these activities, they published the booklet *30 Simple Energy Protocols You Can Do to Save The Earth* in 1991, establishing their operations as part of a general interest in environmentalism (EMA 1991). These were soon followed by work on energy policy and connections with political leaders, such as the then-Vice President Al Gore and the Russian leader, Mikhael Gorbachev.

While these arguments focus on the media's power to influence the brain-print, material considerations have been a substantial part of the EMA's activities from early on. In 1995, the EMA published the first online resource on environmental solutions and practice for the media sector (EMA 2017b). The focus on the mitigation of the material role of production had not been discussed much, if at all, at that time in the industry. If the EMA is a central early node in the emerging network, this document provided vital connective material for its establishment. In doing so, it acted as a founding version of the obligatory passage point (OPP) required for the network. It is a constitutive presence in the maintenance of the network with NGO participants such as Friends of the Earth and Greenpeace, as well as industry organizations like the National Geographic and Disney's Environmentality division becoming integrated through its centralizing presence. While this is very much an EMA environmental management network, these other organizations form networks on their own, as Law has suggested all actors of a network do, and thus, one could study the Disney Environmentality network, for example, and trace its connections with producers, suppliers, NGOs, unions, and so forth to compare how this network's operations differ from the EMA.

This, however, is not part of the agenda for now. What interests us most here is the discursive elements – the connective material – used by the EMA to consolidate its network. The main ordering discourses present in the construction of the network are based on two approaches to the material base: grounded and transferable materialities. *Grounded materialities* indicate the role of material considerations – non-mediatic materialities, such as minerals and toxins as well as production resources, such as electricity and wood – in production practice. These materialities are translated into rhetorical form through policy and practice documents developed for media productions. For example, the EMA's solutions and practice document hosts a key role for these grounded materialities by discussing issues such as recycling and resource use as well as contemporary science on toxicity and pollution. While these are framed from the perspective of their significance for media management, the centrality they hold in designing management strategies testifies to the extent to which material considerations can influence the network. While areas such as recycling and toxins are not covered in much depth, addressed as they are in customary terms to the environmental literature of the time, they still have the power to dictate areas that matter for this particular network. Barring the use of plastic bottles designed for single use from set, for example, is a material saving that impacts positively on the emissions of a production, but it is also a management decision that influences catering decisions and ultimately even the budget of the shoot. As we will see below, these can be framed in multiple ways as a cost-saving mechanism or an impediment to the efficient running of the shoot. Both approaches can reveal some of the embedded politicized inclinations in their rhetoric and can be analysed to uncover some of the vested interests coordinating development of environmentally sustainable production practice. Thus, any discussion of resource practice can be considered as instances of material dictating organizational practice – it is up to the analyst to decipher the particular balance between eco- and anthropocentric rhetoric integrated into these ordering discourses.

Transactional materialities comprise the second type of discursive response to the material base, which is a term used to indicate instances where translations of the material base are used to entice external parties to join the network. Such instances could include, for example, encouraging a producer to initiate sustainable practices on set, or encouraging an organization to join the activities of the EMA. In practice, these are explicit networking activities designed to extend the network to include new participants. Such activities help to construct the profile and prestige of the network and contribute to validating its existence. The translations of the material base become capital in transactions between different parts of the network, hence leading to the EMA documents showcasing a range of these materialities. A clear starting point for these activities comes from phrases such as 'EMA launches its Membership Program to educate and inspire people to live sustainably' (EMA 2017a). While the programme is based on a range of instructions to cut back on consumptive practices and support the green economy, and the rhetoric is relatively basic in tone, it indicates that the principal KPIs of the EMA consist of networking activities focusing on transferring knowledge about environmental sustainability.

While these areas do not focus on the footprint of the media sector, but instead on material mitigation through brainprinting, they indicate the general orientation of the EMA network's organizing discourses. Key to these transferable materialities is the use of celebrity endorsements and award galas. Thus, comments such as 'The EMA President Debbie Levin Presents Josie Maran with Eco Award at Fifth Annual Official First Ladies Luncheon' (ibid.) can be read for their networking properties. The use of recognition of environmental actions in these events transfers the sustainable actions of Josie Maran into a commodity generating PR for both the star and the EMA. While there is an environmental material basis for these actions, here they are transformed into mechanisms to enhance the visibility of the organization. Other ones conflate NGOs and resource politics: 'EMA attends and supports DIGDEEP's Waterless Pool Party in L.A. to raise awareness for water conservation during California's historic drought' (ibid.). The focus on water shortages provides an important material consideration to address, but to do this through a pool party, no matter how waterless, seems more of a gimmick than anything else.

These activities arguably consolidate the normalization of these resource-intensive activities, even as they are used as networking material for environmental purposes. Other activities use social media partnerships to provoke environmental awareness while connecting the organization with brands associated with dubious sustainability practices: 'The EMA supports Just Label It's #ConcealOrReveal day of action on social media' (ibid.). Finally, transactional materialities also include campaigns linked to high profile television shows and their sustainability work: 'Board Member Darby Stanchfield launches our exciting new "Don't Be Plastic" program, installing a hydration station in the makeup trailer for Darby's show *Scandal*, partnering with EMA Corporate Board Member, Star Waggons. *Scandal* is now saving over a 1000 plastic bottles per week on their production' (ibid.). Here, petrochemicals and harmful substances like polycarbonates are translated into corporate PR, even as they demonstrate means of reducing emissions for the sector. While the use of transactional materialities for networking purposes is arguably too close to greenwashing, the logic of this rhetoric fits the KPIs of an organization like the EMA. The glitz and glamour – arguably all with a sizeable footprint in their own right – mould grounded materialities into the rhetoric of the Hollywood film business. This is a practical choice on part of the EMA as galas and social media hashtags fit into a populist pragmatist strategy that makes use of the core competencies of the industry for which they work.

Both types of materialities – the ability of materiality to shape practice (grounded materialities) and its use as part of networking activities (transactional materialities) – often combine in reality. This is the case as management of the industry – both production and organizational management – is based on transactional exchanges, and grounded materialities will need to be incorporated – translated – into vocabulary that matters for industry KPIs. Examples of the EMA activity can explain this. In 2003, the EMA partnered with the Consumer Electronics Association to provide the first-ever environmental symposium on hybrid technologies for the media. This is a significant venture where grounded materialities – developing innovations for production practice, utilizing new renewable resources to replace more traditional sources – feed into corporate transactions.

The use of certifications is another transactional translation that is prolifically used by the EMA. For this, they partnered with the third party green certifier Green Seal in 2009 to recognize programmes that implement sustainable practices into production planning. Certificates are a good example of combining the materialities, as they act as organized translations of grounded materialities into production practices as well as providing visual signifiers of belonging to the network (transactional materialities). Productions aiming to gain the certificate will need to meet certain material goals to be eligible, where eligibility is predicated on meeting material requirements set by external environmental associations. The role of grounded materialities tends to be much more central in shaping policy as certification has to give the impression of scientific veracity, at the very least, to be a credible mechanism. This goes for the Green Seal certificate as they implement a credit structure based on self-assessment using calculators. Each reported act contributes to a points tally that must reach 75 points to gain the certificate. In addition, each production must meet mandatory requirements which include the following:

1 Implemented a production-wide environmental plan, statement and mission.
2 Created an action plan for energy conservation, water conservation and waste management.
3 Created an action plan for socially and environmentally responsible material sourcing for the duration of the production.
4 Made known the production's socially and environmentally responsible practices for vendors.
5 Recycled all materials accepted by a local recycler.
6 Implemented and communicated a strict no-idling policy for all vehicles on set.
7 Determined carbon footprint of current production.
8 Set contains no materials sourced from endangered species, illegally harvested wood or other restricted wood products. (EMA 2017b)

These actions do not generate any points for the report, but are obligatory tasks for a successful application. Assessed areas include a range of grounded materialities including Energy and Emissions, Water Conservation and Management, Waste and Reduction Management, Transportation, Catering and Craft Services, Purchasing and Set Materials, Wardrobe and Makeup, Filming, Office Operations, Education and Messaging, Innovation, as well as Aspirational Items. The highest total of 7 points can be awarded for using renewable energy generated on set whereas carbon offsets obtain 2 points. This suggests that grounded materialities are more significant than transactional materialities in gaining the certification as offsetting is a particularly problematic instance of transactionality where the tangible material savings from on-set actions, for example, are displaced into a financial exchange. Another instance of funnelling attention to concrete on-set practices comes from the 10 points awarded if a production commissions a site visit (ibid.). Accountability based on tangible grounded materialities is thus emphasized as a key priority for participants aiming to participate in the network.

As we have suggested, grounded and transactional materialities tend to combine in organizational practice and can seem somewhat arbitrary (are green certificates transactional as they work largely as PR exercises, but are, simultaneously, based on an extensive verification system?) It is worth distinguishing between them as the extent to which these networks are based on eco- or anthropocentric logic can be deciphered based on this, with grounded materialities emphasizing the former, transactional ones the latter. In some of the later chapters, this distinction will be made through quantitative analysis of organizational rhetoric, but here, they can be traced through qualitative statements. In many of the activities of a consulting organization like the EMA, the emphasis is on transactional networking. Certainly, when working with large brands like Nike, transactionality tends to overtake the material base as a priority. However, in many instances, this relationship is much more complicated.

For example, the EMA's provision of a list of suppliers who meet their sustainability standards makes use of both transactional and grounded materialities. To give a few examples, companies such as The Expendables Recycling provide low-cost excess materials for set construction, Conservatree does purely recycled paper, Green Truck Catering emphasizes sustainable food, Hollywood Rentals specializes in biodiesel generators. These are all examples of building a media management network through transactional materialities. At the same time, grounded materialities are a considerable influence on these production decisions even to the extent of dominating them. If the EMA acts as a network OPP between production companies and suppliers, its role may be transactional, but its ultimate role is to facilitate connections between a production company and specialists in a particular field of grounded materiality. Consequently, this will reverberate in the network, ensuring a more grounded constitution for its environmental activities.

Biodiesel provides a good example of how grounded materialities are adopted by production companies and facilitated by the transactional role of the EMA. The translational mechanisms of EMA designate points for biodiesel as it has better greenhouse gas benefits compared to conventional diesel fuel. Two will be awarded for productions using B20 fuel (20 per cent of total concentration is biodiesel), four for those with a higher concentration. The EMA recommends Standard Biofuels to provide B20 as it 'represents a good balance of cost, emissions, cold-weather performance, materials compatibility, and ability to act as a solvent' (ibid.). Thus, the production company commissioning standard biofuels for biodiesel integrates these grounded materialities into its production practice, allowing material considerations to dictate production management and budgeting. For the EMA, this comprises a transactional boost that meets its role as an OPP for the industry.

Discursive materiality and the EMA

Grounded and transactional materialities operate as key indicators of the eco- and anthropocentric directions of the EMA's operational parameters. By focusing on the organization's discursive practices, we can start to understand how these translational processes operate as power structures governing the network the EMA maintains. They can be explored by evaluating the constitution of its ordering discourses from qualitative and quantitative perspectives. Analysis of key words comprises a quantitative indicator to evaluate how materiality is negotiated in these discourses. They can give a snapshot of the rhetorical slant of a document by identifying some of the key priorities of the organization producing it, though they are by no means scientifically infallible data. Through this, they provide a means to identify the main directions of the discourse, and thus, the particular eco- or anthropocentric perspectives adopted by particular aspects of the network. I call these directions the managerial and the environmental – descriptions that roughly coincide with the transactional and the grounded respectively – to indicate the ways the EMA orders the connective content of its network.

Here, I focus on a document called *Green Rider*, published by the EMA (2015). This document lays out some of the key priorities of the organization, including all the relevant material areas that impact film and television production. A survey of the key word frequency of the document reveals some of the overarching priorities of its ordering discourses. The key words include the following (as well as frequency of appearance):

- production – 20
- trailer – 19
- studio – 10
- water – 10
- environmental – 9
- footprint – 9
- carbon – 8
- water bottles – 4
- solar-powered – 4
- private jets – 3
- first-class – 3
- serviced apartment – 3

The terms can be categorized according to the managerial or the environmental category based on the role they afford to material agency. Thus, terms like production, trailer, studio, private jets, first-class, or serviced apartments are managerial terms that indicate more anthropocentric considerations in the organization of production. Terms such as water, environmental, footprint, carbon, water bottles, solar-powered can be considered more specifically ecocentric as they can be seen to refer to explicit environmental measurements or grounded materialities. Of course, the allocation of terms in either category is not absolute as the process relies on a degree of subjective judgement, but such categorizations can act as indicators of some of the underlying priorities of each document. In addition, the implications are not so obvious in the case of an environmentalist consultancy, such as the EMA, whose mandate it is to be environmental, but they will be more prominent when we explore organizations like the BBC who have an interest in environmental sustainability but do not consider it a KPI.

The key word frequencies in *Green Rider* show that the managerial and the environmental areas are more or less balanced. This fits with the mandate of the document which is to explore environmental and carbon savings through contract negotiations between talent, studio and the representation. The introduction of environmental sustainability into contract negotiations works as a translational process where environmentally-aware talent is expected to encourage the studio to adopt environmental measures. This is best exemplified by specific instances of rhetoric prioritizing the role of talent in the network. The document initially presents sustainable practice as a means to minimize energy use and decrease the waste generated on a production. While reducing the carbon footprint of a production is seen as important, the pitching of this process to the studio is more revealing: 'any such reduction is in the best interests of the studio due to cost and environmental savings' (ibid.: 1). In addition, as these are to be achieved by key talent requesting them as part of their contract, 'in exchange, the studio should be asked to donate a percentage of money saved to a cause or nonprofit of the artist's choice, thereby creating positive PR for all involved' (ibid.: 1). The balancing of discourse includes all the key rhetorical devices used by sustainability organizations in attempting to entice collaboration throughout the industry. These are the key ordering discourses that structure networks: environmental initiatives, material concerns, economic logic, and publicity. As we will see, they are repeated with some frequency in most of the documents analysed in this book.

When approaching these discourses, it is worth remembering that the emphasis on costs is a pragmatic approach that acknowledges the industry's priorities as a business and calibrates its actions accordingly. While economic language and contractual specifics arguably reterritorialize material considerations, they are done in order to bring materiality back onto the discussion table. This approach has a significant impact on the ways in which materialities are translated. For example, one of the key achievements listed by the EMA is the savings of $400,000 on the 2014 production, *The Amazing Spiderman 2*. These savings were achieved by instigating comprehensive sustainability measures on set. Yet, the ways environmental sustainability is framed matters, as any heavy emphasis on financial savings can be seen to constitute an act of reterritorialization, and priorization of anthropo-centric managerial goals by the EMA. Grounded and transactional materi-alities can be used to analyse this balance, for example, in the organization's critique of production gifts: 'both start and wrap gifts are inherently wasteful' (ibid.: 1), they state. Such strategies can be seen as ways to address some of the most egregious expendable activities of the industry, cutting back on the grounded materialities that feed into such procedures. Instead, the aim is to 'seek a monetary donation through one of the studio's charitable causes' (ibid.: 1).

Such mitigation acts are largely transactional and do not cut back on the direct emissions of a production, but both strategies provide viable means to ensure environmental sustainability is being considered appropriately. Yet, even as the gifts provided at wraps are 'inherently wasteful', these events, often hugely consumptive affairs, are otherwise left without much criticism. Similar problems are evident with travel, for example. When it comes to private jets, the document suggests they are expensive for the studio and incredibly wasteful sources of greenhouse gas emissions: 'Alternative travel arrangements are required. Talent would be offered an entire row in first class for each flight during production' (ibid.: 1). This is complemented by the next sentence: 'NOTE: This airfare accommodation is only for produc-tion, not for press junkets and other engagements, which require talent to fly continuously on short notice' (ibid.: 1). Again, the problematic behavioural patterns of the industry are largely left unchecked by these mitigation processes.

While emissions from grounded materialities are mitigated by switching emphasis to transactional means – talent demands environmental measures; the studio may be able to save money or they generate positive PR – it is clear that the emphasis is only on mitigation, not full-scale address of the underlying problems. Similarly, the framing of demands on the talent show how business realities and material considerations collide: if talent is demanding a 'triple-wide, or double-wide trailer, seek to downsize'. Instructions are to demand

> an eco-fitted trailer with energy-efficient lighting, water filtration, recycling bins, and potentially, solar panels. Request that no water bottles be stocked on board. Additionally, demand that all generators, lights, and A/C units be shut off when the trailer is empty. Going further, demand the use of solar-powered generators to power trailers in the base camp.
>
> (ibid.: 2)

All the instructions rely on material expenses which can curtail the envir-onmental costs of a production, but simultaneously, these recommendations continue to normalize hyper-consumptive practices as industry protocol. The framing of these policies reveals a key problem: while the emphasis on stars to demand changes in management protocol is arguably realistic in the case of well-known 'green celebrities' like George Clooney and Leonardo DiCaprio, the rhetoric used by the EMA only solidifies the worst excesses of Hollywood. Grounded materialities are here subsumed into negotiations where potential emissions reductions become transactional capital. While these can be useful in curbing emissions, they also emphasize the ways grounded materialities can be translated to solidify the industry's emphasis on a culture of excess and accumulation.

At the same time, it is worth remembering that the Green Seal application asks for a considerable level of information on material expenditure to attain the certificate. Detailed information on grounded materialities from the production supply chain is provided, for example, when pertaining to cleaning supplies being non-toxic and environmentally friendly; food sourced from local and/or organic sources, and in-season wherever possible, and leftover untouched food items must be donated to a local food bank; cups, plates and/or utensils being reusable, biodegradable and/or compostable; paper consisting of at least 30 per cent recycled materials; trailers powered by a solar-powered generator capable of replacing standard diesel generators or if necessary, by a generator employing biodiesel at a concentration of no less than B20 (EMA 2017b). These are key concerns exemplifying how grounded materialities establish best practice guidelines for productions, which can also have a more productive role to play than simply act as transactional materialities used to construct business or PR opportunities.

Industrial realities

The evaluation of the balance between eco- and anthropocentric rhetoric is thus a contested process and forms a reality that may appear contradictory, and even fragmented. The complex constitution of networks is both an unavoidable and necessary reality. Law reminds us that networks are about enactment and performance. Different agents assemble to 'enact a set of practices that make a more or less precarious reality' (Law 2007: 13). In discussing how organizational management networks operate in large laboratories, Law suggests that a network needs to provide an impression of simultaneous creative flexibility and administrative control for it to operate sufficiently. The management of the laboratory setting requires innovative scientists to act independently to meet the demands of their performance while bureaucratic procedures ensure protocols are followed and respect for administrative processes maintained. In tandem, they facilitate circumstances where the freedom and flexibility essential to experimental work are valued while experiments are conducted within the strictures of scientific laws. Creativity and bureaucracy are ordering discourses in this network as they facilitate appropriately dynamic roles for the different actors while they maintain the coherency and purpose of the network – to provide innovative work that fulfils the standards of the organization.

The combination of creativity and careful management comprises the reality of Hollywood film production as well where environmental areas are perceived as operational concerns, as the Responsible Media Forum argues. Thus, they do not fit with its KPIs which tend to be premised on efficient management and cost reductions. For EMA, the network they have created is a conscious ongoing performance to assert environmental goals. The EMA's role is to create an alternative reality where environmental areas become meaningful to the extent that they can be considered strategic concerns. As we have suggested, this can be achieved by framing the adoption of environmental strategies differently either by resorting to economic or management rhetoric, or alternatively by emphasizing pragmatic and practical considerations. But we must also remember that the EMA is not an environmental consultant for the industry but rather a publicity organization. They do not produce these protocols, but instead work to promote networking among diverse industry stakeholders. The balance between creative decisions and administrative process is thus not about innovating with new production methods but with thinking how to get the industry invested in environmental sustainability. It is about the need to make environmentalism a more visible part of the industry, and thus, the reality that the EMA oversees is one predicated on publicity and image management.

As the 'reality' of the EMA network highlights publicity and political recognition, transactional materialities play a key role here while grounded materialities are largely ignored. 'Agents' such as Robert Redford and festivals including Sundance act as other nodes in the network, forming transactional exchanges and acting according to their core competencies, which are to do with generating wide publicity and debate over these issues. Other actions conducted by the EMA include hosting briefings on the state of the environment with Robert Kennedy, meetings with the Clinton administration and Al Gore and receiving the President's Environment and Conservation Award from George Bush in 1991. NGOs such as Environment California, Americans Against Fracking, SoCal Edison and California Green Communities as well as consultancies such as Green Globe (a hotel certification operation), the Environmental Protection Agency Green Power Locator, the California Integrated Waste Management Board, and Green Planet Productions (who provide PAs trained in green production) also act as key nodes in creating public awareness and extending this network into communities outside of the media industry. While these affiliations do not contribute to the development of environmental media policy, they show that the core competencies of the EMA are in claiming political legitimacy by establishing associations with a range of external organizations. To these ends, they suggest that there has been 'a massive show of support from the entertainment industry [on] a major expansion of environmental oversight throughout the physical production process, with the placement of new environmental stewards, sustainable production managers and sustainability coordinators at various studios' (EMA 2017a). The EMA plays a significant part in getting these ideas to the forefront of debates on the relationship between environment and media though this is not the type of networking that traditionally provides transactional capital for the organization.

The difference between these networking activities becomes more obvious when we compare the EMA's approach with that of the Producers Guild of America's (PGA) Green section. The PGA was one of the first organizations to start acknowledging the footprint of film production. In their 2014 document, *Going Green and Saving Green: A Cost-Benefit Analysis of Sustainable Filmmaking*, written by eco-supervisor Emellie O'Brien (2014), the PGA provides a step-by-step analysis of how green production methods can save productions money. They use a combination of practical and economic strategies as an enticement for production companies, indicating how these two areas act as powerful ordering discourses for the PGA network. They suggest that 'the myth that it costs more to "go green" on set is currently plaguing the production of film and television' (ibid.: 2). In response, the document explores a range of methods to 'show how pro-environmental measures can translate into budgetary savings for productions' (ibid.: 2). The target group here consists of producers and senior management instead of the EMA's explicit external networking. O'Brien suggests that 'when resistance by decision-makers occurs, it is often due to complaints over cost' (ibid.: 2). Responding to these obstacles is a matter of leadership within organizations and ensuring an 'Eco manager/Eco-supervisor' can oversee communications and the coordination of sustainable practice.

If the EMA's reality was one of public image management, ensuring that environmental areas are addressed by the industry, the PGA's is one concerned with negotiating environmental actions in relation to both production budgets and timescales. Translational tools are essential for them as well but here the role they play is much more centred on grounded materialities. The aim of the PGA network is to communicate 'cost savings percentages calculated from production accounting records as well as current market prices of respective vendors' (ibid.: 2) with the aim that this feeds into the environmental reports of these companies and eventually to modification of consumptive behaviour. While the emphasis on costs operates in a similar way to transactional materialities, the focus on targeting producers through practical savings generated by sustainability measures makes this less about translating materialities into costs (whereby the rhetoric would likely try to downplay costs from new environmental strategies) but more about ensuring that costs are translated into materialities (whereby the act of emphasizing material savings will lead to reductions in costs). These debates frame perceptions of environmental sustainability as a financial or managerial cost as the problem – they emphasize the ways impressions of costs plague production development, and how complaints over costs prohibit adoption of these practices. The difference is small but significant as this discursive angle prioritizes the ability of materialities to have agency over the constitution of the network which, when communicated in this form to industry participants, can translate both into material savings and as a means to anticipate and overcome resistance by managerial components of the network.

The brief comparison between the EMA and the PGA shows the distinctive approaches of different environmental organizations. The EMA is much more PR-focused whereas the PGA emphasizes the role of materialities in productions. The repeated use of terms like partnering, consulting, hosting, and participating in the EMA documents implies that the networks they foster operate on a transactional basis. In contrast, the PGA is much more focused on internal industry communications. Simultaneously, these two networks combine to create a sustainability network for the film and television industry. The network comprises a particular reality with its own sets of rules and limitations on what is practical and feasible to achieve within it. The precarious reality discussed by Law comprises, for this network, circumstances where industry resistance is anticipated and thus different ordering discourses are required to consolidate the significance and feasibility of environmental sustainability strategies.

Bibliography

Barad, Karen. 2007. *Meeting the Universe Half-way: Quantum Physics and the Entanglement of Matter*. Durham, NC: Duke University Press.

Bennett, Jane. 2009. *Vibrant Matter: A Political Ecology of Things*. Durham, NC: Duke University Press.

Bourgine, Paul and Stewart, John. 2004. 'Autopoiesis and cognition', *Artificial Life*, 10: 327–345.

Braun, Bruce and Whatmore, Sarah. 2010. *Political Matter: Technoscience, Democracy and Public Life*. Minneapolis, MN: University of Minnesota Press.

Coole, Diane and Frost, Samantha. 2010. *New Materialisms: Ontology, Agency, Politics*. Durham, NC: Duke University Press.

Cubitt, Sean. 2017. *Finite Media: Environmental Implications of Digital Technologies*. Durham, NC: Duke University Press.

EMA. 1991. *30 Simple Energy Protocols You Can Do to Save the Earth*. Beverly Hills, CA: EMA.

EMA. 2015. *Green Rider*. Beverly Hills, CA: EMA.

EMA. 2017a. 'Historical timeline'. Beverly Hills, CA: EMA. Available at: www.green4ema.org/historical-timeline/ (accessed 11 November 2017).

EMA. 2017b. *Green Seal Application*. Beverly Hills, CA: EMA. Available at: www.green4ema.org/ema-green-seal/ (accessed 11 November 2017).

Foucault, Michel. 1981. *The Order of Discourse*. New York: Routledge.

Fuller, Matthew. 2005. *Media Ecologies: Materialist Energies in Art and Technoculture*. Cambridge, MA: The MIT Press.

Guattari, Félix, 2000. *The Three Ecologies*. New York: Continuum.

Iovino, Serenela and Oppermann, Serpil. 2012. 'Material ecocriticism: materiality, agency and models of narrativity', *Ecozon*, 3(1): 75–91.

Law, John. 1992. 'Notes on the theory of the actor-network: ordering, strategy and heterogeneity', *Systems Practice*, 5: 379–393.

Law, John. 2007. 'Actor Network Theory and material semiotics', available at: www. heterogeneities.net/publications/Law2007ANTandMaterialSemiotics.pdf (accessed 11 November 2017).

Lundby, Knut. 2009. *Mediatization: Concepts, Changes, Consequences*. New York: Peter Lang.

Luppicini, Rocci. 2014. 'Illuminating the dark side of the internet with actor-network theory: an integrative review of current cybercrime research', *Global Media Journal*, 7(1): 35–49.

Maxwell, Richard and Miller, Toby. 2017. 'Greening cultural policy'. *International Journal of Cultural Policy*, 23(2): 174–185.

McLuhan, Marshall. *Understanding Media: The Extensions of Man*. New York: McGraw-Hill.

Mukherjee, Rahul. 2016. 'Mediating infrastructures: (Im)mobile toxicity and cell antenna publics', in Walker, Janet and Starosielski, Nicole (eds) *Sustainable Media*. New York: Routledge, pp. 95–113.

O'Brien, Emellie. 2014. *Going Green and Saving Green: A Cost-Benefit Analysis of Sustainable Film-making*. New York: Producers Guild of America Green.

Parikka, Jussi. 2012. *What Is Media Archaeology?* Cambridge: Polity.

Parikka, Jussi. 2013. 'Green media times: Friedrich Kittler and ecological media history', in *Archiv für Mediengeschichte*. Munich: Wilhelm Fink Verlag69–78.

Parikka, Jussi. 2015. *A Geology of Media*. Durham, NC: Duke University Press.

Postman, Neil. 2005. *Amusing Ourselves to Death*. Harmondsworth: Penguin.

Räikkönen, Timo. 2011. 'The greening of work: how green is green enough?' *Nordic Journal of Working Life Studies*, 1(1): 117–133.

Slayton, Joel. 2005. 'Foreword', in Fuller, Matthew. 2005. *Media Ecologies: Materialist Energies in Art and Technoculture*. Cambridge, MA: The MIT Press.

Thompson, John B. 1995. *Communication and Social Context*. Cambridge: Polity.

Walker, Janet and Starosielski, Nicole. 2017. *Sustainable Media*. New York: Routledge.

3 Media policy in actor networks

Introduction

Environmental management of the media operates as a complex collection of agencies that can be studied through the discursive practices of media organizations. These discursive practices and communication strategies form the primary analytical material for this book. Thus, this is not a scientific study of the physical processes that have been developed to curtail the industry's footprint. Nor is it an ethnographic production-based exploration of the industry. Instead, the aim is to uncover the complex agencies and power relations that take place in management and policy frameworks. These frameworks organize the communication of values and priorities for the media industry and leave material traces in the form of the ordering discourses of particular management networks that can be used to trace the complex negotiations that advance or restrict the adoption of environmental measures. These traces are the policy and corporate documents that we analyse in this study.

The work of Cubitt, Starosielski, Vaughan and Gabrys focuses on the underlying material processes with considerable clarity, zeroing in on the ways these processes are understood and incorporated into policy. They allow us to understand the power structures involved in facilitating dialogue between eco- and anthropocentric approaches in ways that may eventually consolidate into media policy. This emphasis is especially relevant as terminological and conceptual complications often prohibit industrial adaptation of sustainability strategies across different industrial sectors. The lack of shared goals and common vocabulary to integrate the environmental sciences and media production practice is at the centre of the challenges to developing media-specific strategies, as a study on the environmental impact of the mobile media infrastructure states:

> In addition to specialised terminology, different actors in product value chains may have difficulties understanding the overall lifecycle of the product their operations are connected to. Especially background processes, such as energy, chemicals and raw material production, can be difficult to identify.
>
> (Dahlbo *et al.* 2013: 851, 859)

The ways the industry struggles to adhere to scientific veracity while ensuring effective communications across the network make it absolutely essential to find appropriate vocabulary and establish mutually productive modes of communication to ensure grounded materialities are thoroughly integrated into the network.

Analyzing media materiality is not only about evaluating production practice but also about understanding the wider regulatory environment which encourages and curtails these practices. Media producers have to operate within political and economic regimes that shape much of what is feasible within the industry. Regulations, standards, legislation, procedures and agreements are all both discursive *and* material practices that translate material considerations into transactional elements, albeit to meet very specific anthropocentric goals. As we have suggested, tracing the roles of grounded and transactional materialities in documents published by media organizations can be productive for revealing key power structures and agencies that shape the constitution of a network. In approaching such policy documents as maps for tracing eco- and anthropocentric concerns in the practice of environmental media management, this analysis draws on Marijke De Valck's (2007) use of documents as significant organizing material for film festival networks. In charting film festivals as actor networks, De Valck suggests that to understand the power relations of film festivals, pro-grammes, brochures, company and film promotions, specialist line-ups, and so on influence visibility and, thus, power in a festival network. The posi-tioning of an individual film in a festival line-up affords it specific levels of visibility and prestige. Being associated not only with certain festivals but other films that circulate in these networks boosts the chances of individual films gaining visibility and distribution. Accumulation of value occurs with the flow of texts through various festivals which changes the properties of a film. It is then not only the agentic qualities of artists and their products that matter here but a wide range of other associations that hold power over the ability of ideas and objects to traverse through these networks.

Uncovering these associations in the environmental media network requires that we explore the complex regulatory and policy environment that surrounds the industry. To do so, we will have to cover international envir-onmental law, specific media regulation frameworks, domestic policies for the environment and for cultural production, the internal practices of media organizations as well as the work of environmental consultancies focusing on the media. We view this network as a dynamic evolving reality, where our aim is not to propose future guidelines for the industry or produce protocols to overcome its contemporary obstacles in adopting environmental sustain-ability. Instead, the aim is to chart the dynamic systems that characterize this 'media ecology' as a network of diverse agentic materialities. The point is to provide an ecological understanding of the network as a dynamic system 'in which any one part is always multiply connected, acting by virtue of these connections and always variable, so that it can be regarded as a pattern rather than simply an object' (Fuller 2005: 4). Approaching the actor-network as an assemblage of different agentic components coordinated through discursive mechanisms provides a means to map patterns in which environmental concerns have been incorporated into the media industry.

Environmental management of the media

Richard Maxwell and Toby Miller's *Greening the Media* (2012) provides a productive starting point for this analysis, especially as it lays the groundwork for evaluating the industry's regulatory infrastructures and management processes. The study can be considered as an initiation for this field as its coverage ranges from more obvious areas of the media's footprint such as petrochemicals used for film stock to the hazardous components of mobile technologies. Furthermore, the discussion incorporates relevant case studies from Google's data centres to Hollywood studios to make the case for a material understanding of the media. The role of digital communications, for example, is phrased in a way that accentuates the urgency of addressing it from the perspective of both regulators and industry:

> Greenpeace estimates that if the cloud were a country, it would be the fifth largest energy consumer in the world. With a few exceptions, the major data centers are doing very little to wean themselves from coal, gas, nuclear, and other dirty or dangerous sources of electricity. To do so requires retrofitting existing plants or moving operations to within reach of renewable energy.
>
> (ibid.: 4)

The authors argue that data farms are huge consumers of electricity and contribute considerably to the 2 per cent of global CO_2 emissions which Greenpeace estimates as the footprint of ICT (some estimate this at 3 per cent). Furthermore, they address regulations curtailing the disassembly of media devices and its associated detoxification and recycling processes including the 1992 Basel Convention on the Control of Transboundary Movements of Hazardous Wastes and Their Disposal as well as Waste Energy and Electricity Emissions (WEEE) and Restrictions on Hazardous Substances (RoHS). Such regulations prohibit the international transportation of hazardous material and act as essential guidelines for the industry even as the effectiveness of international regulatory frameworks is also frequently questioned by the authors. China is one of the destinations of the global market in recycled devices from the 'affluent West' and the government has established exten-sive regulations banning this trade. Still, this has not restricted sites like Guangzhou from becoming one of the largest sites for illegal shipping and disassembly of these devices. The exponential increase of mobile users in the domestic market has only contributed to the environmental and human costs at sites like this and others all over the country. Similar areas are also opening up in other parts of the Global South to accommodate the considerable human and environmental costs of the media industry in all parts of society. This poses a moral dilemma and a fundamental challenge that is essentially 'a global problem that sits at the core of capitalist development' (Maxwell and Miller, cited in Kooijman 2013).

For Maxwell and Miller, these concerns are not only about the media industry, but ones that concern the essential fabric of contemporary society:

> Can we transform the system from one built on values of growth, abundance, and consumerism to one based on sufficiency and sustainability? In addition, can the media's grand aesthetic and political claims to being sources of pleasure and of knowledge stand up to serious scrutiny of their indubitable costs? Because simply by consuming, audiences are endangering the world they occupy.
>
> (ibid.)

Clearly, media need to be regulated, but outside of ICT laws, few comprehensive regulations pertaining specifically to media production have been developed. As the regulatory environment is lacking, for Maxwell and Miller, one way to ensure that companies contribute to managing their own footprint is to highlight the role of accountants. They argue that 'green citizens of the future ... must keep in check the managerial, human-centered tendencies of CBA and risk managerialism that fracture a holistic understanding of the relationship of media technologies to the environment' (2012: 159). The authors are well aware of the utopianism of this suggestion, especially for an industry that relies on increasing levels of efficiency. Cost benefit analysis and risk aversion are management strategies that often prioritize the anthropocentric logic of transactional discourse that, as suggested above, amounts to a reterritorialization of the material base of the network. Indeed, they suggest that accountability faces opposition from many levels of the contemporary mediascape, ranging from legislators to producers, from practitioners to consumers. Yet, the inclusion of more ecocentric accounting mechanisms would comprise particularly productive coordinates for a functioning network.

Intriguingly, neither the Environmental Media Association (EMA) nor the Producers Guild of America (PGA) have included the accountant as a key priority for consolidating environmental sustainability as part of the practice of the industry, but have instead underlined the roles of celebrities and producers, who meet the organizational priorities of their networks more clearly. Accordingly, the construction of their networks is premised on different imperatives than those where the accountant might be the most obvious obligatory passage point (OPP). While this emphasizes the diversity of different media management networks, it also shows how the environment occupies varying levels of importance in the different networks of which the media industry consists. It rarely acts as a value unto itself and operates more as a means to generate a set of interest groups based on the particular priorities of each network.

Corporate social responsibility (CSR)

To take this discussion of diverse management strategies further, we now move to explore corporate social responsibility (CSR) strategies for media companies. Here, we focus on the Media CSR Forum, a consultancy working with a range of international companies on understanding the industry's responsibilities. Since 2001, they have organized workshops to discuss industry perceptions of corporate responsibility. From these events have emerged a set of documents that place environmental considerations in the wider spectrum of CSR. The Forum's work provides an important addition to the environmental accounting perspective emphasized above as it addresses multiple aspects of the production and consumption chain of the media industries. Simultaneously, CSR is a conceptually problematic field in a similar way to sustainable development as it has often been hijacked by economic concerns that underscore transactional benefits with minimal environmental gains. Strand and Freeman (2015) argue that there has historically been confusion over the applicability and overlap between sustainability and CSR, and indeed, they share many conceptual roots. The European Commission, for example, has outlined CSR as 'the responsibility of companies on their impact on society' (European Commission 2017). Much as the definition of sustainability in many policy debates fits with the green capitalist aims of the Brundtland Commission, CSR is a concept that has its roots firmly in corporate activity, a fact that limits any attempt to comprehend it as an absolute societal or environmental good. As it is premised on facilitating business activity it needs to be evaluated for the extent it values corporate logic over environmental progress, and how these manifest in the balance of the eco- and the anthropocentric rhetoric in its ordering discourses.

Keeping these complications in mind, we approach the Media CSR Forum as an OPP in its particular environmental media management network. Its publications provide us with an overview of how the company constructs the network and how eco- and anthropocentric discourses facilitate its development. Since 2005, the Forum has conducted three studies at five-year intervals to assess the salient CSR priorities of companies operating in the media sector. The data for the studies come from its members, consisting of a broad section of the media, including publishers, digital companies, broadcasters, local and national newspapers, internet service providers, as well as satellite and cable TV channels. During discussions, the Forum asked these companies to identify CSR issues that pertain to the industry (KPMG 2005). This approach meant that the industry would have to identify its own value systems and position its operations in a material environment. As the questions focused on CSR in general, most of the findings emphasized communicative values such as diversity, literacy, creative independence, transparency and impartiality, all areas considered specific to the media sector. Other common CSR issues included a range of technological areas such as IP, copyright, data protection, and finally, the promotion of sustainable development. Significantly, the last category involves issues which are not specific to the media sector, and apply to most industries, areas, such as supply chain, corporate governance, community investment, and environmental management. It is clear from these discussions that environmental sustainability is not a particular priority for the sector, a process that has continued in future reports.

In 2008 (Media CSR Forum 2008), the emphasis was on social and environmental issues as key promotional goals, and by 2012 (Media CSR Forum 2012), these would both be integrated into the term 'sustainable development'. As the term sustainable development is mobilized as a sort of catch-all term for environmental issues and CSR, it also functions as a conceptual tool to merge all areas of cultural, political, societal and environmental development under a singular heading. On top of its problematic connotations as a metaphor for green capitalism, the use of sustainable development as a catch-all synonym for CSR obfuscates the role of environmental concerns. Furthermore, the Media CSR Forum has outlined some of the key organizational management concerns of the media industry into three categories, which tell us of the relative importance of an issue for the company: material, strategic and operational priorities:

- A *material issue* is financially significant over the short to medium term, i.e. it has the potential to affect a key financial indicator, e.g. profits or revenue, by around 5 per cent or more within a two-year time period.
- A *strategic issue* has the potential to significantly affect the ability of the company to deliver its strategy in the medium to long term.
- An *operational issue* matters for other reasons – internal, reputational, efficiency – but is neither material nor strategic. Under normal circumstances, it does not represent a significant threat to the company (Media CSR Forum, 2013a: 5).

The first two of these indicators emphasize areas that are of fundamental significance to contemporary media corporations. A material issue for a company like Google would be an increase in energy tariffs, which would necessitate reorganization of its resourcing. A strategic issue may concern a film studio like Twentieth Century Fox ascertaining ownership of a franchise to ensure it can keep producing films in the X-Men series, for example. Thus, content production or the provision of cheaper bandwidth and storage as the foci of competitive strategies for media companies – they comprise the core competencies of these companies and provide their major sources of revenue. However, environmental considerations would fall under operational concerns as they would not, in most cases with media, comprise a considerable threat to either content production or technological areas like cloud services.

This point is made even more complex when we factor in that when environmental concerns do matter, they do so in ways that relate to the specific core competencies of different media sectors. For example, when discussing the environmental strategies of publishing companies, the Media CSR Forum suggests:

> [The] media has a relatively low impact on the environment relative to its scale. Water, waste and other emissions are not significant to a media company's performance ... However, the sourcing and printing of paper can present a major cost with companies exposed to volatility and security of supply questions.
>
> (ibid.: 14)

The sourcing of pulp for a publishing company such as Reed Elsevier, for example, would be a material concern due to the role these grounded materialities play in their core competencies. The strategy fits with the arguments of the Forum in highlighting the ways material issues impact investment in sustainability – if it influences a core area of the company's performance, it attains the level of strategic concern. Yet, for a film production company, for example, the sourcing of pulp would not form a strategic concern. In fact, the simple act of recycling paper may be sufficient here, and thus this would comprise an operational concern for them.

The investments of different media companies also vary considerably. For example, a company like Sky Broadcasting (Sky 2017), carbon-neutral since 2006, leads the industry in many ways as it has invested in low carbon technology and accentuates employee education to cut down on its emissions. One of the key performance indicators (KPIs) for a broadcasting company is enhanced reach through using the most cost-effective technology. Its sustainability goals will be different from Elsevier in that recycling would only be a small part of their focus. Other important areas for Sky emerge from office management, the use of low emissions filming technology, HR behaviour, and cutbacks on travel. Yet, credibility and managing a positive reputation are also key drivers for these companies. Reputational capital is a key incentive for Sky, arguably to navigate some of the more problematic public relations disasters of its parent company, News International. Thus, they emphasize environmental sustainability in all areas of operations and focus not only on their core competencies but on managing the entire footprint of their operations. These concern areas such as supply chain integrity, whereby companies are 'expected to act transparently and ethically when buying goods and services' (Media CSR Forum 2013a: 15). The Forum notes that companies have increasingly started to concentrate on the 'particular issues that matter to them. This choice is highly individual, usually based on whether the issue will affect – directly or indirectly – the company's performance in the years ahead' (ibid.: 7). For companies like Sky, these would be areas that provide reputational gains through making this part of their corporate operations more ethical. But regardless, environmental sustainability is not a strategic or monetary issue for a company like Sky.

Indeed, for most companies, evaluating the scope of the footprint does not rank high on their list of priorities:

> [M]ost media companies are not energy intensive. Even when including the whole supply chain, energy and carbon costs are an immaterially small fraction of the total value. A changing climate will certainly affect media companies' operations, but their business models would seem to be sufficiently flexible to deal with it.
>
> (Media CSR Forum 2008: 14)

From the perspective of a company working to generate strategic or material value, the level of emissions and cost savings do not amount to anything of substance that would provide commercial or strategic advantage for them. As these arguments are often rhetorically contrasted with the role of heavy industry for whom environmental areas are often legal – thus strategic – and economic – thus material – it is not surprising that the Media CSR Forum suggests that the focus on the footprint may not be the most productive approach to take for identifying CSR targets for the industry:

> So far the sector has been treated as 'low impact' in terms of its direct sustainability effects. But surely its biggest impacts are intellectual and psychological. Relatively speaking, media companies make up only a small fraction of global business and resource usage, yet they command an inordinately high share of 'voice', a voice that is both powerful and pervasive. Media content, in all its forms, touches the daily lives of almost every human being on the planet. Its footprint may be modest, but its 'brainprint' is enormous.
>
> (ibid.: 15)

For them, 'the unique contribution of the media sector' (Media CSR Forum 2013b: 7) is in shaping public debate, changing behaviours and promoting sustainable lifestyles. They argue that the focus on the carbon footprint is misleading in that it does not provide a productive understanding of how the media can contribute to society. To confirm this assertion, they conducted a study of the sustainability reports and relevant sections of the annual reports of ten major media companies from 2005 to 2011. The analysis discovered that discussion of content accounted for 27–40 per cent of the companies' reporting efforts, with 'the majority of the space taken by the more traditional topics of suppliers, community, environment and employees' (ibid.: 16). This is an intriguing suggestion as it indicates that, from a CSR perspective, the media companies are already engaged in more than enough work on their footprint. For the Media CSR Forum, 'The evidence suggests – not surprisingly – that media companies are much more comfortable with metrics in areas where more readily quantifiable data is available; environment and employment leading the way' (ibid.: 16).

In contrast to these tangible measurements, the ways that content leads to social change is much less clear and is more difficult to estimate. One way to start developing such mechanisms is to conceptualize of ways to inspire audiences, to amplify content with socially and environmentally beneficial perspectives, to reduce harmful communications regarding lifestyle choices and even to develop ways to normalize environmentalist arguments. Measuring how these succeed in imprinting on audiences and influencing societal progress is very much work in progress currently and is now emerging as a key theme for consulting companies. In the perspectives emphasizing brain-printing, transactional materialities are the main asset of value. While grounded materialities play a part in some of the earlier scoping work, their relevance in establishing concrete CSR policy is at best tangential. Yet, in contrast to this emphasis on brainprinting, I will argue that focusing on measuring carbon emissions and resource use is a key imperative for media companies.

Cultural inflections

One of the areas overlooked by these consulting organizations is the role that cultural differences play in both the willingness of an organization to adopt environmental sustainability and the shape these strategies take. Albarran (2005) reminds us in his study of organizational media management that corporate cultures are one of the key influences for the types of policy and managerial practice established in a given company. Yet, such areas are often absent from the work of consultancies like the EMA or the Media CSR Forum. This is not entirely surprising as the focus of these organizations tends to be on the political economy of the media infrastructure and its regulatory management. Yet, it is also clear that culture is a discursive tool that imposes a set of values on these strategies and inflicts how a particular node channels communications through the network, instilling imperatives and value systems for a particular actor network. Here, the various nodes – the companies – operate through the OPPs – sustainability initiatives that meet relevant standards – based on the particular cultural constitutions they adhere to. The EMA works more obviously on the basis of PR than the production cultures approach taken by the PGA. In contrast, the Media CSR Forum emphasizes the political economy underlying decisions in corporate management. The constitution of ordering discourses for these different networks relies on discursive material that must satisfy both the company KPIs and their stakeholders, especially in terms of the cultural values they hold.

Other cultural formations also contribute to the ordering discourses that structure the operations of these networks. These factors include key areas such as regional, national, societal and ideological cultural formations, including the specific regulatory and policy frameworks coordinating the media industry in, for example, national contexts. Such areas can be productive for understanding how a network is structured based on infrastructural concerns that influence the constitution of organizational cultures. Yet, we must be mindful of prioritizing culture over the materialities essential to structuring environmental networks. Otherwise, this discussion would only reinforce anthropocentric conceptions by over-writing – reterritorializing – such materialities. As we have suggested, these explorations must focus on the ways culture works as a discursive ordering mechanism and evaluate the constitutive factors that shape the networks based on these cultural inflections. For example, both the EMA and the PGA need to be considered in the context of Hollywood as the particularities of this industry influence how these networks come to be. Thus, for the EMA, its celebrity and political connections play a role in the transactional form its network takes. For the PGA, the scale of Hollywood production cultures makes its protocols both an aspiration and a complication for other industries to try to emulate.

At the same time, we must remember that corporate cultures are often highly self-conscious constructed notions that vary by context. By this we mean that there are elements in each organization that emerge out of their specific cultural settings, but also that organizations consciously brand themselves according to cultural and economic imperatives that meet their KPIs. Thus, the PGA both reflects values that make sense in the context of the US film industry but also self-consciously recreates its leadership for a global media environment. Thus, as we will see in the analysis below, much of the work done on sustainable film policy in other contexts tend to emulate the PGA model. At the same time, the Media CSR Forum, rebranded as the Responsible Media Forum in 2017, highlights its role as a cross-media operator that provides strategies for considering the business benefits of environmentally sustainable media. The culture of the Forum consists of highlighting its operations as an international leader capable of keeping the economic realities of media organizations at the forefront. They state that 'in the fast-changing world of media, we are flexible and business-led but always open to input' (Responsible Media Forum 2017). This characterizes well its value systems premised on global operations, and the necessity of adapting to the flexibility and speed of international business. Clearly, then, we must not overlook culture as a key factor underpinning the environmental management of the media. This is especially the case as culture – as both a manifestation of anthropocentric worldviews, and a potential avenue for instigating more complex reflections on how to integrate ecocentric thinking into corporate management – sits at the heart of the debate over the role of the media industries as a material presence and a communicator and facilitator of ideas over environmental degradation and progress.

Case studies

The analysis to follow will highlight culture by exploring case studies of a number of key documents published by relevant media organizations from a range of countries in the European media industries. By identifying key words from each document, we can start to chart the main priorities of each organization and their perspective on environmental management. This is complemented by qualitative analysis of rhetorical statements with the aim of consolidating a more comprehensive idea of how each particular network forms and connects with other networks. From this, the book will address the following four research areas:

1 Accounting is a key paradigm for the industry but exploring both management and practice indicates the complexity of professional roles that impact this conduct.
2 Contextual differences are of vital importance in understanding both regulations and practice of environmental media and often shape the adoption of such activities.
3 The different qualities and quantities of rhetoric translate into concrete material that reveals the eco- and anthropocentric perspectives of key areas – such as KPIs and stakeholders – and of each of the networks.
4 Content is considered king by the industry but this obfuscates self-reflection over how content is produced and what its ethical and economic underpinnings are.

These areas are addressed in the following order. Chapter 4 will focus on regulation on the intragovernmental EU level as well as within the Nordic countries, which will, from now on, form one of the key areas of study. From there, we chart the sustainability initiatives that have been established in the European media industries with examples drawn from production and consulting organizations operating in different national contexts. The discussion is divided into four types of networks based on the extent and functionality of these networks. The first concerns the film and television media management network in the UK where environmental practice is relatively well established. The discussion will evaluate the state of the industry and how it has arrived at a situation where environmental practice is increasingly becoming the norm. The second will focus on continental European efforts to establish environmental policies and formulate common rhetoric for them. This section – a network in formation – will focus on the infrastructure – the regulatory environment – in which media organizations operate to analyse how networks can emerge or are constricted by contextual factors. I will then move to examine different parts of the Nordic media industries with Chapter 6 focusing on different sectors: broadcasting, publishing and film and television. This is a network largely without infrastructure. The first part explores the complications of constructing regulatory frameworks and policy for an industry that is not considered a priority for environmental legislation. The second part focuses on different sectors of the media to highlight diverse practices mobilized by individual production companies to overcome the lack of regulatory oversight, and suggests alternative ways that sustainability can be integrated into the fabric of the industry.

Bibliography

Albarran, Alan (ed.) 2005. *Handbook of Media Management and Economics*. New York: Routledge.

Dahlbo, Helena, Koskela, Sirkka, Pihkola, Hanna, Nors, Minna, Federley, Maija, and SeppäläJyri. 2013. 'Comparison of different normalised LCIA results and their feasibility in communication', *International Journal of Life Cycle Assessment*, 18(4): 850–860.

De Valck, Marijke. 2007. *Film Festivals: From European Geopolitics to Global Cinephilia*. Amsterdam: Amsterdam University Press.

European Commission. 2017. *Corporate Social Responsibility*. Available at: http://ec.europa.eu/growth/industry/corporate-social-responsibility_en (accessed 11 November 2017).

Fuller, Matthew. 2005. *Media Ecologies: Materialist Energies in Art and Technoculture*. Cambridge, MA: The MIT Press.

Kooijman, Jaap. 2013. 'Greening media studies: an interview with Richard Maxwell and Toby Miller', *Necsus*, Spring.

KPMG. 2005. *The Media CSR Forum*. London: KPMG.

Maxwell, Richard and Miller, Toby. 2012. *Greening the Media*. Oxford: Oxford University Press.

Media CSR Forum. 2008. *Mapping the Landscape: CSR Issues for the Media Sector*. London: Media CSR Forum Secretariat.

Media CSR Forum. 2012. *The Media CSR Forum Activity Report*. London: Media CSR Forum Secretariat.

Media CSR Forum. 2013a. *Does It Matter: Material, Strategic or Operational?*London: Media CSR Forum Secretariat.

Media CSR Forum. 2013b. *Mirror or Movers: Framing the Debate about Media Content*. London: Carnstone.

Media CSR Forum. 2015. *Mirrors or Movers*. London: The Media CSR Forum Secretariat.

Responsible Media Forum. 2017. 'Partnering for a sustainable sector'. Available at: https://responsiblemediaforum.org/home (accessed 11 November 2017).

Sky. 2017. *Environmental Policy*. Available at: http://s3-eu-west-1.amazonaws.com/sky group-sky-static/documents/bigger-picture/policies-2016/environment-policy-july-2015.pdf (accessed 11 November 2017).

Strand, Robert and Freeman, R. 2015. 'Scandinavian cooperative advantage: the theory and practice of stakeholder engagement in Scandinavia', *Journal of Business Ethics*, 127(1): 65–85.

4 Material rhetoric

Introduction

The adoption of green practices across different media sectors has met with both resistance and encouragement, involving debates on establishing policy for the industry to critiquing the use of environmental causes as publicity ventures. Academics tend to perceive a lot of these ventures as greenwashing. For example, Lipschutz and Rowe (2005) argue that the industry's efforts to date do not have any significant impact on climate goals and only work as minimal efforts to showcase the ethical side of the industry. Such strategies work well for companies needing to generate positive impressions of corporate value with various stakeholders but do not constitute appropriate activity on mitigating their own role. For example, the Media CSR Forum's rationalization of the role of environmental sustainability for the sector would seem to validate this as 'most media companies are not energy intensive. Even when including the whole supply chain, energy and carbon costs are an immaterially small fraction of the total value' (2013: 14). Thus, when companies do turn their attention to sustainability practices, this is not so much concerned with taking care of an area posing a potential risk for the company, but rather to take advantage of a potentially useful area of corporate social responsibility (CSR) public relations (PR). According to Lipschutz and Rowe, the establishment of sustainability goals has much more to do with business than the environment. Yet, others argue for a more pragmatic take on how sustainability and industry can be intertwined. Bottrill, Liverman and Boykoff in a study of environmental practices in the music industry suggest:

> While carbon-based industry interests have often been the ones criticized for defending their political economic interests over social and environmental concerns, the debate has certainly become more textured and nuanced in recent years. Yet, the business sector remains widely seen as pivotal to efforts to decarbonize society through both voluntary actions as well as through regulatory compliance with government programmes that seek to cap and reduce emissions.
>
> (2010: 2)

The authors argue that the business community is an essential partner in implementing environmental actions. They can lead on a range of essential tactics for environmental management including the following:

> reducing energy use and increased consumer confidence through the perception of environmentally responsible branding, pre-emptive response to the possibility of regulatory controls, business leaders or shareholders seeing a moral imperative to act and the comparative advantage in being the business forerunner, gaining experience, innovating early, and setting best practice standards that may influence policy.
>
> (ibid.: 2)

The description summarizes the key imperatives for developing comprehensive CSR strategies that meet the goals of sustainable development policies. At the same time, the statement suggests that the material base for media production needs to be integrated as part of these sustainability management strategies. The first part of the statement starts out from the material base of the network – the grounded materialities – and indicates how they need to be translated into transactional materialities as they are funnelled into PR activities. Other parts of the statement cover the regulatory environment, ethical necessities of combining moral considerations with business practice, and finally the necessity of establishing the type of network seen with the Environmental Media Association (EMA) and the Media CSR Forum, one that is based on innovation and leadership on a future priority area. These arguments are heavily reliant on the anthropocentric connotations of transactional materialities, as befits the explicit business management slant of the rhetoric.

In contrast to this business-focused logic, many of the academic works, such as Lipschutz and Rowe (2005), discussed above, propose best practice that highlights a more idealistic sense of ethical and practical integration. They adopt a clearly ecocentric perspective to the balance of grounded and transactional materialities that would, most likely, critique the above rhetoric as green washing. Walker and Starosielski refer to this balancing act as entanglement: 'To be entangled is not simply to be intertwined with another, as in the joining of separate entities, but to lack an independent self-contained existence. Existence is not an individual affair' (2016: 39). The description makes a lot of sense in that it captures the ways in which all business practices will invariably be 'grounded' to the material base. Transactional materialities may be the way for organizational management to talk to its stakeholders about these priorities but these discussions have to be based on material, grounded, realities. Yet, the idea of entanglement mostly works on an idealistic, descriptive level, and it is certainly not a rhetorical tool used by the industry to describe its actions. It is necessary to adopt a sense of practical realism here as 'holistic and integrated approaches promise to tackle and balance everything with everything at the same time. However, the risk is that in the end they amount only to fine rhetoric on principles – and little action' (Hey 2009: 18). Simultaneously, these more holistic perspectives must be adopted as useful conceptual frameworks to evaluate industry practices, especially in terms of any sectors that considerably deviate from ecocentric principles. They are an essential critical checking mechanism to ensure the business priorities of the media do not translate grounded materialities to transactional values without due consideration.

To evaluate how this balance is negotiated in media management strategies, we must pay attention to the ways the rhetoric of sustainability is conceptualized and actualized in strategy and policy statements. Corporate documents, as we have suggested, act as material actualizations of ordering discourses that control the ways a particular network – as well as its placement within other networks – is conceptualized. For such a network to function as an ecocentric operation, all actors would need to be operating on an equal level with power imbalances at a minimum. This is not the reality of the industry, however, and thus, it is necessary for these networks to create their own realities, as Law (2007) has suggested. It is these realities that this chapter aims to interrogate through textual analysis of key policy and strategy documents from central environmental media organizations. I will especially focus on uncovering the ways eco- and anthropocentric emphases emerge in these documents through the use of ordering discourses premised on grounded and transactional materialities, and especially how the human is not 'the originator or the locus of these systems but one component in it' (Fuller 2005: 2).

While it is vital to keep environmentalist considerations at the centre of this study, I will adopt a more pragmatic approach that views both environmental critique and industry environmental rhetoric critically. For one thing, even more idealistic or ideological framings of ecocentric rhetoric often have anthropocentric underpinnings, for example, as we have seen with the industry's emphasis on PR or cost-saving strategies. As we have frequently suggested, for the contemporary reality of the media industry, environmental concerns are simply marginal. The preference for the economic is echoed across the media industry and is often framed in the form of what I call a reality discourse for the industry. This is based on the principles of economic competition and gain and the act of using sustainability as a way to open alternative revenue streams for companies. Due to these inherent imbalances, it is necessary to keep this reality in mind when approaching the often aspirational rhetoric of these organizations as well as the holistically oriented academic rhetoric that tries to envision more environmentally sustainable ways of conceptualizing the media's ecological presence.

The material

The material for this analysis of discursive practices consists of two types of publications through which the industry communicates about its environmental responsibilities. These consist of, first, promotional material and publicity documents and, second, environmental strategies and corporate reports published by media organizations. To provide the chapter and the book as a whole with a solid analytical basis, the documents I include here consist of material from a range of cultural contexts. But for this and Chapter 5, they mostly come from the UK which has been an early adopter of substantial initiatives for environmental management of the media. These include both regulatory strategies by cultural organizations such as Arts Council England and Creative Scotland (covered in Chapter 5) as well as incentives by key public and private media organizations such as the British Academy for Film and Television Arts (BAFTA) and the British Broadcasting Corporation (BBC) (covered in Chapter 5). I will also include material from the Environmental Media Association (EMA) and the Producers Guild of America (PGA) as a comparison base. As we have already outlined, these organizations have instigated environmental policy for the screen media industry and their strategies are significant not only for paving the way for environmental policy for the sector, but also for setting best practice and quality standards for the industry. Another set of participants analysed here includes non-governmental organization (NGOs) and consulting organizations – mediators between industry and regulators – including Julie's Bicycle and the Media CSR Forum. To widen the scope of analysis, I also focus on documentation from the German consultancy Green Film Shooting and the international industry organization Cine-regio as the media industry has relatively similar priorities, despite cultural and contextual variations in the size of the industries and the regulatory frameworks that oversee them. Analysis of these documents will reveal not only connections and power relations between them, but also the connections these organizational networks have to the wider material context in which the media are produced.

The analysis of the ways the different discourses bind the networks together, and of the ways they incorporate material and human agency, focuses on discursive framings. The study of framing can unpack the ways these documents act as selective versions that pick 'some aspects of a perceived reality and make them more salient in a communicating text, in such a way as to promote a particular problem definition … of the item described' (Entman 1993: 52). This established means of content analysis will allow me to evaluate the ways these discursive practices shape the directions of their particular network operations. Accordingly, the documents I address frame environmental activity in particular ways and reveal some of the key ideological positions these nodes – the companies in question – adopt inside the networks. If we consider the actions of the EMA, for example, the use of frames to connect with a range of celebrity and political players positions environmentalism as a worthwhile popular cause. The EMA sets the agenda for the need to consider the footprint of the industry but framing the majority of its activities as being concerned with fundraising or galas paints a different picture. This comprises a transactional use of the material base and communicates two messages. First, that the network capitalizes on the power of the media in distributing environmental ideas across popular culture. Second, that the high profile of these beneficial CSR actions generates reputational capital – especially important for an industry premised on popular attention. The construction of discursive positions consists of the 'selection of a restricted number of thematically related attributes' (McCombs *et al.* 1997: 37), such as celebrity culture and political ambitions, or environmentalism and social work, to provide connections for this particular actor network. The combination of these elements through most of the EMA's publicity frames its environmental activities as both the right thing to do as well as an economically profitable strategy.

Environmental sustainability in the UK

How do some of the other media organizations communicate their environmental responsibilities? How do these organizations frame the salient issues of the debate – and thus connect with other nodes in the network? The obvious way to start to trace the evolution of an environmental management network would be to start at the top. This would most likely include regulatory oversight in the form of governmental or intergovernmental bodies legislating for a particular sector. But, as we have already established, such regulations do not exist in any comprehensive form for the media. To navigate this problem, I will start this discussion by focusing on developments in the UK film and television industries. At the time of writing, the UK has established arguably the most advanced environmental initiatives for the media industry. Several organizations are devoted to ensuring that the footprint of the sector is managed, ranging from cultural organizations like the Arts Council England, the BBC, BAFTA and Creative Scotland, to private companies like Dogwoof Productions and *The Guardian*. Their strategies range from consulting on environmental production methods, developing best practice, enhancing means for voluntary participation, and allocations of production funding. The operations in this context provide a comparative, and potentially even a transferable model for other similar schemes, especially as the sector has been able to consolidate both voluntary participation and mandatory requirements as part of the everyday fabric of the industry.

Why has the UK media industry been so keen to take up these initiatives and act as a global instigator of environmental strategies? How have they managed to get both policy-makers and practitioners on board? One of the key reasons is certainly the Climate Change Act of 2005 which requires the UK to cut its emissions by 80 per cent by 2050. As film and television production, for example, often relies on public funding, they have to abide by the general parameters of this legislation. In practice, the limited scale of emissions from the media sector has restricted the extent to which they feature in these overall climate goals, and thus, the extent to which domestic legislation applies to them. The majority of areas that these regulatory frameworks influence are to do with the energy infrastructure and organizational management concerning recycling and procurement protocols, for example. We will focus on such infrastructural considerations later in this study, but the key point to take from here is the limited impact that governmental regulations have on environmental sustainability policy for the media industry.

Even so, we can note a range of examples of top-down environmental incentivizing in financing arts and cultural production. Two of the major funding organizations for the arts and culture sector, the Arts Council England and Creative Scotland, have made it mandatory for recipients of financial support to report on CO_2 emissions. To receive any financing from these organizations, producers must use a carbon calculator and provide data on their environmental practice. This is a viable alternative strategy to, for example, the PGA model that only *encourages* users to adopt sustainable practices. As such, it proposes a considerable alteration to the structures under which environmental practice has operated internationally. One of the reasons for the development of these models arguably comes from the UK's CRC Energy Efficiency Scheme (CRC). This is a mandatory carbon emissions reduction scheme that applies to large non-energy-intensive organizations in the public and private sectors. At the time of its inception in 2007, these choices were made based on a consultation from 2006 where mandatory, instead of voluntary, reporting mechanisms received considerable support. The principles of the scheme have now affected culture and the arts, yet the CRC has since been decommissioned and will finish in 2019. While this does not mean that the mandatory requirements of the Arts Council of England or Creative Scotland will be decommissioned, it does show the precarity of policy aimed at sustainability, especially when environmental protocols start to have an influence on all levels of corporate practice.

If the Arts Council England and Creative Scotland comprise particularly important nodes for setting the agenda for environmental media sustainability in the UK, other key stakeholders operate as more specific nodes of the network. These include the NGO Julie's Bicycle, a consulting company that has played a key role in developing sustainable strategies and calculators for the arts. In addition, the British Standards Institution, an NGO with a Royal Charter, developed the BS 8909, the British Standard for sustainability in film production, in 2011. Both companies have conducted extensive work on assessing the environmental impact of screen production and used these data to develop strategies specific to the sector. They have collaborated on designing the sustainability policies of the British Film Institute (BFI) in its five-year plan *Film Forever 2010–2015*. This is a strategy to outline sustainable film production practices for the following resource-intensive areas: energy, refurbishments, waste management and information platforms. Targets such as sustainable buildings, excessive travel, recycling on set and renewable energy are common-sense areas that can be calculated and communicated to various stakeholders in ways that ensure environmentally sustainable media production makes sense. As part of the environmental management network in the UK, both BS 8909 and *Film Forever* act as translations of grounded materialities that consolidate connections with production practice and the material base, ensuring that they are also translated across the different nodes of the network.

While these are some of the key nodes in the UK environmental media network, a particularly powerful OPP in this context is the carbon calculator Albert, developed in a collaboration by the Arts Council of England and the BFI. The initiative is led by the BAFTA Albert Consortium, established in 2011, and consisting of 13 major broadcasters, including the BBC, ITV, SKY, and Channel 4, as well as production companies, such as Warner Bros, NBC Universal, Fremantle Media, and Endemol. The consortium is a good example of a media management network that shares an understanding of an industrial reality that here consists of the benefits to be gained from adopting sustainable strategies. Kevin Price, the CEO of BAFTA, outlines the key strategic principles of the initiative as follows:

> Individually, the creative industries have done much to promote and embed sustainable practices. Nevertheless, by sharing opportunities, challenges and aspirations across the sector we stand to achieve much more. Our challenges and solutions are by no means unique and I firmly believe a collaborative approach to be a catalyst for greater progress.
>
> (BAFTA 2017)

The necessity of conceptualizing sustainability as a network comes through strongly here. Industry leadership and guidance are important in bringing the sector together, but at the same time, the network needs a distinct identity that is able to distinguish, for example, the film and television industries as a separate sector with specific concerns that arise out of its particular material conditions. Constructing this network is thus premised on ordering discourses that emerge from a much more grounded sense of responsibility compared to the EMA, for example. While the transactional relations between the network participants are used to promote sustainable practice, this is not framed as a business or PR opportunity, but a catalyst 'for progress', at least in the rhetoric.

This arguably ecocentric rhetoric pervades the Albert report, *We Know How to Make Greener Telly: Now Let's Get Down to Business* (BAFTA 2015). The title indicates that the focus is on leadership among media peers but in such a way as to divest it of hierarchy or formality. By engaging potential participants on a voluntary basis 'the consortium is creating and sharing "gold standard" sustainability production stories with expectant audiences and industry peers who are keen to learn' (ibid.: 2). The offer to join the network is based on the provision of transferable strategies that are projected to make participants become aware of emergent practices but also to cater for new audience demographics. The statement – 'a group on a mission' (ibid.: 2) – makes it clear that the consortium aims to educate and enhance awareness of sustainable practice but also to take part in regulatory debates over the industry's footprint. It is the organization's mandate to contribute to 'developing all the required skills, providing the necessary tools and materials and building the essential relationships' (ibid.: 3).

While there is a strong sense of organizational management here, the discursive angle of the report emphasizes the agentic role of grounded materialities. The actor network in this case is premised on material connections which draw on emissions data based on 1400 statements submitted for the project during its three years. These include 38 million KWh of electricity, 69,000,000 air miles, 1,800,000 litres of water and an additional 80,000 tonnes of other carbon emissions. The totals are all intended as translations of material concerns into large sums which give an impression of the scope of the material base and the expenditures the industry has. These rhetorical strategies promote the rationale for environmental sustainability and are boosted by comparisons to solidify the connections of the network. For example, the document compares the industry total to 20 Space Shuttle launches, or alternatively, they note that the production of one hour of television amounts to 9.4 tonnes of carbon, which is the equivalent of the annual emissions for an individual in the UK. The document also acknowledges that the real footprint may be considerably higher as it now includes only information that has been voluntarily submitted. If the industry was mandated to report all of its emissions, a much more striking picture of the field's significance would certainly emerge, the report seems to suggest.

While these translational statistics consolidate the necessity of the network, they also contribute to normalizing sustainable practice as part of the everyday reality of media production by indicating how pervasive the industry's footprint actually is. Furthermore, these are consolidated by framing sustainable activities in ways that promote convenience by repeating terms like 'easier' or 'possible'. Other phrases include those designed to give the impression of accumulation in industry adoption, whereby 'the network has grown' and participants would be 'supplementing an already healthy bank' of knowledge (ibid,: 3). These provide a feasible incentive to adopt sustainable practices ahead of the curve before they are normalized as standard operating practice. Finally, the availability of case studies throughout this document provides industry benchmarks for curbing emissions and ensuring grounded materialities maintain a presence in the ordering discourses of the network. These examples provide adaptable models and set standards for the network by indicating how grounded materialities can be adopted across these productions:

- *Wonders of the Monsoon*: Using international cameramen the team saved upwards of 100 tonnes of CO_2 by avoiding baggage transport and crew travel across their six shoots.
- *Stargazing LIVE*: Using waste vegetable oil in their generators saved over 500 kg on carbon on this one-day event.

- *Operation Grand Canyon*: Using solar power for the whole shoot meant the Grand Canyon teams saved some serious carbon and could access locations otherwise off limits.
- *Springwatch*: Champions of low carbon power, the team used small and large renewable generators for the unit base and remote camera set-ups.
- *Trollied*: By quizzing crew on carbon reduction, this series of *Trollied* cut their stage power needs by 50 per cent, sourced pre-used set dressings and helped feed local livestock with surplus fresh produce.
- *Invictus Games*: *Invictus Games* got to grips with sustainability by addressing transport head on; specialist vehicles on-site and putting crew and talent on public transport scored this BBC show some extra points (ibid.: 5–7).

Each one of these programmes is highlighted as examples of Albert-certified production due to their specific material advances. These strategies include sustainability measures in travel, renewables, catering, energy efficiency, and recycling. As earlier, the evaluation of media production KPIs is invariably focused on efficiency and costs with BAFTA translating these into easy-to-understand measurements to make them viable as connecting material for the network. Thus, the document uses carbon equivalents for different emission types and provides estimates in large quantifiable sums such as 100 tonnes of CO_2 and 500 kg of carbon. While these are translational devices, their methods of framing material costs are grounded in environmental realities and use material frames to communicate the impacts instead of cost estimates, for example. Thus, a key question for our analysis is concerned with how the translation process takes place, instead of if it takes place. The distinction is important as media strategies will inevitably reflect their anthropogenic roots, and hence, it is the type and scope of the rhetoric that matter.

The rhetorical strategies of the Albert Report exemplify many of these directions. We have so far identified calls for coherent group activities, addressing producers via examples, suggesting convenience and ease as strategies for adopting sustainability, and presenting data in an understandable format. Other advantages come from accessing locations, involving the crew, championing incentives and scoring points. These are all typical management strategies communicated in a positive tone that emphasizes motivational and aspirational benefits and easy-to-reach goals. Contextualized within the actor network, they can be seen as translations of material constituents of the network into the anthropocentric rhetoric of organizational management. Through this, material elements are integrated into the strategic planning of these media companies. Accordingly, the consortium suggests that leadership initiatives are key to ensuring that 'commissioners understand the potential environmental impact of the programmes they order, and work with production companies to reduce them'. Senior management and producers are identified as key network participants who must ensure that their 'production units are aware of the green options available to them and use them on their programmes' (ibid.: 8). This indicates that sustainability would need to be integrated at the conceptualization stage, factored into all aspects of production management and include 'green power generation and procurement, sustainable set builds, responsible travel and efficient production and post-production technology' (ibid.: 8). These strategies are encouraging in that they explicitly place agentic materialities at the centre of production management. While CBA and other more anthropocentric organizational management strategies will surely play a part in production planning, companies adopting the Albert protocols will have to integrate a much more pronounced ecocentric angle into their strategies.

Paratexts

The Albert documents provide ecocentric strategies for the industry, establishing the tone for other key parts of the UK network, from the BBC to the Arts Council. But what role do such documents occupy in the day-to-day management of the industry? To position these and other documents in the UK network, I use John Caldwell's (2013) concept of 'the para-industry' as a framework for understanding how they operate as vessels for ordering discourses. Caldwell's use of the concept draws on the notion of the paratext, often referred to as an adjacent or additional text to primary media content. Gerard Genette has termed paratexts as texts that prepare us for other texts (1997), whereas Jonathan Grey uses the term an 'airlock to the text' (2010). Both approaches are premised on the existence of a principal text. Paratexts, in comparison, are those texts that exist alongside the main text and contribute and shape our perceptions of it in complex ways.

These sorts of texts often have a promotional orientation that aims to contribute to the audience's interest and understanding of the main text. However, they are not direct promotional material in themselves in the ways that, for example, film trailers are. Paratexts can be industry blogs or making-of documentaries that highlight aspects of a production, often in ways that provide an aura of support for the industrial underpinnings of the main text. The paratext has been studied from the perspective of industry promotional workers (Grainge 2011) and fans (Hills 2015). Both studies focus on the ways the blogs of below-the-line labour or unofficial edits of official content by fans expand the original text even as they challenge some of the industrial contexts or ideological content associated with the original work. This point is significant as the challenges – such as fan fiction or online databases – are often incorporated into the dominant structures of the industries by mechanisms such as intellectual property (IP) negotiations or aspirations on the part of paratextual producers to be acknowledged as an adjacent part of a franchise or a brand.

My focus on policy, production and other industry documents works from a similar paratextual basis. These documents are significant internal statements on values that exist alongside the texts – plays, festivals, performance art, film, etc. – that cultural industries produce. They can be considered as across-the-line paratexts as they talk directly at management and labour, relating concerns that have specific meanings for the performance of their daily practice. Even though these documents occasionally address external stakeholders, they are not the priority of these communications as the key impetus is on ensuring that clear guidelines and aims are set for different parts of the media industries. The publications for an incentive like Albert operate in a paratextual sense as Albert is not part of the core process of getting a television show on air but acts more as a way for the industry to set standards in its own area of operations, which are then communicated internally and externally. As with any paratext, they come laden with meanings and implications for industrial relations that, in this case, provide an invaluable resource to analyse the ways grounded materialities and management coalesce. Thus, by focusing on these intra-industry communications, we get a better understanding of how the industry perceives its responsibilities on the environment, including the type of emphases it places on mitigating environmental harm.

While paratexts are especially useful as sources for industry strategies, John Caldwell has taken the idea further by focusing on what he calls the para-industry:

> I adapt 'para-industry' here less from Genette's view of para-texts (as material 'surrounding' the primary text) than from 'paramilitary' (since Hollywood closely mirrors the production 'outsourcing' and critical analysis of the Iraq invasion to Blackwater-type subcontractors). Hollywood's Blackwaters don't just produce 'content.' They also traffic in the theories, oppositional postures, and analytical justifications that scholars have developed to maintain objectivity and distance.
>
> (2013: 158)

While there is a tinge of conspiratorial collusion in this description, it also describes well many of the tactics used by the industry to promote its values and beneficial roles in society. Environmental policy acts as a verification that the industry is partaking in the general greening of the arts and uses this to contribute to societal welfare. This is a strategy that draws on both NGO operations – often in collaboration with them – and frameworks from the environmental sciences and sociology. Networking is a key aspect of para-industrial strategies as the inclusion of both CSR consultancies and environmental scientists provides justification for the claims made by these media organizations. Yet, I do not adopt the term uncritically, as Caldwell sees it more as a manipulation strategy to encounter potential oppositional views or critical perspectives. My point is not to argue that these documents are produced for such collusive purposes, as instead, as we have suggested, the strategies implemented by BAFTA, for example, can be justifiably considered ecocentric. But to address some of the problems with greenwashing identified earlier, it is worth exploring industry documents with this sense of the para-industry in mind but adapted so as not to dismiss the very real attempts of regulators, managers, producers, as well as above- and below-the-line workers who take a clear environmentalist view of their work.

There are also other reasons to view these documents as para-industrial in orientation. First, they do not coalesce around just a single text as do most of the conventional discussions of paratexts – instead, the focus tends to be on the state of the industry as a whole. Nor are they simply about branding an organization where each individual text contributes to an impression of its core values and identity. Many of these documents are concerned with the strategies and value choices made by the sector as a whole in order to promote an image of itself both internally and externally. They exist in a paratextual relationship to the key texts of the industry which consist of both the content produced by the different companies – its bread and butter – but also a range of promotional material designed to market these core products. Environmental strategies, certainly, have a pronounced promotional factor as they not only communicate values internally but also link to external organizations and NGOs through visible OPPs like Albert. At the same time, the environmental work of companies including the BBC and the BFI do not focus so much on the companies themselves as they do in ensuring industry-wide participation. Thus, statements they produce on environmental policy can be productively considered as para-industrial texts, indicating some of the cultural shifts in sustainable media production for the film and television sector as a whole.

Promotional material

A key part of para-industrial communications – that is, communications that seek to contribute added value to the operations of the industry – is to do with its close connection to image management and branding. Caldwell suggests that these areas of the industry often emerge at times of 'industrial contestation and change' (2017: 45). This is the approach taken by most of the sustainability networks as they seek to offer companies the opportunity to be early adopters of significant considerations that will impact the industry. As sustainability is arguably not a strategic concern, the rhetoric of the sustainability network tends to be somewhat different from the buffer zones of most para-industrial work. Greening Film (www.greeningfilm.com), a website run by the BFI, provides a good example of such para-industrial communications. It is one of the principal means through which the BAFTA consortium recruits and informs partners as well as communicates their ideas to external audiences. Thus, it is a useful resource for identifying the dominant ordering discourses used by the network. While the page is an outward-facing document and can be used for mapping external connections, its main focus is arguably to be the first point of contact for industry participants. Its purpose is not to be a complicated, verifiable scientific document but an easy point of access for interested parties.

The site provides an insightful glimpse into the organization of a successful sustainability network, especially in the ways it presents the organization's three main drivers for its operations. These are the moral, the economic and the legal drivers, which, when seen as part of the actor network, also operate as its main ordering discourses. Ethical concerns are framed as both the ability of films to communicate environmental messages to wide audiences and as competitive incentives for the UK industry, with comparisons to Warner Bros and Fox Film's environmental work. The section talks about links with the Global Action Plan, which is an incentive offering training courses for organizations to improve the sustainability initiatives they may be developing. The Legal section discusses the Climate Change Act and the work of the Department of Energy and Climate Change (DECC) and also provides information on a range of EU regulations that link to filming regulations in the UK. The argument here is to anticipate the enforcement of regulations on film shoots and establish a competitive basis for the UK film sector. Finally, the economic section focuses on the benefits of environmental measures in reducing costs through implementing material savings. The information provided under these main headings is not particularly novel in its own right but the placement of this section on the front page is very revealing. The economic driver is represented by an image of a pound sign made up of green spheres, which contrasts with the representative images for the ethical and the legal sections, which have no particular relevance to their theme and only use clichéd visuals of spectators. The commercial incentive is clearly centralized and holds the most eye-catching position amidst the content of the page.

One of the documents available from the site is the Greening the Screen toolkit, a comparatively short text providing the 'essential' tools a company needs to implement sustainability. A key focus for documents such as this is to frame sustainable production as an accessible and achievable activity and thus, it is not difficult to identify instances where economic and legal concerns take central importance. These are especially prominent in the emphasis on business considerations which provide an anthropocentric transactional approach to integrating agentic materialities into the network:

> Give yourself a competitive edge – show you're ahead of the curve by reducing your impact now. The more companies get serious about green issues – auditing their performance, thinking ahead and investing in new systems and technologies, the more money they find they can save.
>
> (Greening Film 2011: 3)

The use of transactional materialities positions the argumentative slant of the organization in the realm of anthropocentric discourse, further emphasized by arguments for legal incentives (under the heading 'The Legal Driver' on the organization website) premised on the UK industry staying ahead of other similar incentives, and that this network remains a competitive one: 'Even if your company is reluctant to act now, it makes sense to be aware of coming changes in environmental regulation at local, national and European level. Planning ahead reduces the cost of disruption when new legal Standards are introduced' (ibid.). The economic and legal arguments position sustainability as a transactional means for media companies to innovate ahead of competition.

However, these arguments would not work in isolation and require validation from more pragmatic and ethical concerns. Thus, the Greening Film site suggests that 'CO_2 emissions are staggering with film and television productions in London alone producing 125,000 tonnes of CO_2 each year' (ibid.). Here, grounded materialities play an alarmist counterpoint to the economic and legal imperatives and establish footprinting as a credible and necessary part of production practice. This is especially important as sustainability measures tend to be perceived as add-ons and are thus met with a considerable degree of hesitation. Such arguments are used to build a case for the pragmatic necessity of adapting sustainable production methods. Motivational rhetoric is mobilized throughout the document to these ends, linking managerial considerations with sustainability. This can be seen in statements such as 'Going green is easy!' (ibid.: 2), which makes use of marketing rhetoric to frame sustainability as an exciting strategy, not a legal chore.

The document continues by confirming the considerable benefits of sustainable practice: 'Simple changes can make a big difference, and could just improve your quality of life' (ibid.: 2). The motivational rhetoric here is complemented by more pragmatic argumentation, presumably to complete the investment – to close the sale – from individuals motivated enough to open the document, but who may still require a final push: 'Environmental issues are serious. We are all individually responsible for reducing our carbon footprint in the workplace and at home' (ibid.: 2). These rhetorical framings position these practices as an ethical task, but also as a part of everyday life and – ideally – production practice. The legal, the economic and the ethical drivers are finally complemented by rhetorical work that emphasizes a pragmatic/practical driver. The use of motivational rhetoric and lifestyle politics contribute to the standardization of green practices in the industry: 'Set an example. Help prompt practical action across the industry' (ibid.: 2). The use of these drivers aims to consolidate a reality where environmental production practices become thoroughly normalized.

These moral, business, legislative and lifestyle choices constitute specific repertoires that form the material consolidating the ordering discourses. These are translational devices that provide substantial information on the agenda underlying their formulation – and thus on their general ideological context. In identifying these linguistic repertoires, I draw on the work of the Institute of Public Policy Research (IPPR), who identify three groups of climate change repertoires in the UK media. These include an 'alarmist' repertoire, which is fundamentally pessimistic of the status quo and the ability of society to respond to these challenges. In addition, two types of 'optimistic' repertoires include ones that assume 'it'll be alright' and a more pragmatic set of repertoires that assume 'it'll be alright as long as we do something' (IPPR 2007: 4–8). This is not the place to debate the veracity and applicability of these concepts as, instead, they are more productively seen as a suggestion for the necessity of categorizing rhetoric into functioning indicators that can be used for discourse analysis. Such categorizations are necessary as these linguistic systems indicate a normative 'common sense' that applies to specific stakeholder and audience groups – that is, they reflect the particular reality of a group and help to normalize its practices. For the environmental management of the media, these consist of economic, legal, ethical and pragmatic repertoires. These repertoires outline the key industry approaches to sustainability management and thus establish the common-sense parameters for how specific stakeholders approach sustainability as a strategy and, thus, indicate a set of patterns and conventions that can be unpacked to understand power relations in the media management network.

These repertoires are in use on most UK sustainable media sites. For example, the BBC Sustainability site employs them in their Corporate Responsibility strategy and explains their approach to sustainability in the following terms:

- drive innovation in our core business in order to lead the broadcasting industry in sustainable production;
- continually work to reduce the impacts of our operations, through targets which address our environmental footprint and save the BBC money;
- inspire our staff to exemplify sustainable behaviour in all that they do. (BBC 2017)

Business and practical management take a central role here as the BBC emphasizes integrating innovative sustainability plans into its core business practices by exploring how they can be used to cut costs. This is an important distinction as the impetus here is not on implementing sustainable measures in the least expensive way but instead to design plans to cut back on the current expenses of the organization ('drive innovation'). The volume of space devoted to cost benefits indicates that this is the key incentive used to entice the industry to adopt sustainability measures. These strategies show some of the underlying tensions concerning sustainability in the industry, especially when a publicly funded body like the BBC also resorts to prioritizing economics. The reality for the industry, at least as can be gathered from this document, emphasizes the need to anticipate resistance to integrating sustainability into its operations. Through this, the heavy emphasis on economic repertoires – signifying the transactional, anthropocentric approach to environmental sustainability – emerges as the dominant normalized perspective to adopting environmental incentives.

Similar approaches can be found in the BFI documentation for BS 8909, which frames this management system standard as a tool to 'help the industry stay ahead of legislation both in terms of the environment and social responsibility, and aid the UK film industry by being a pioneer in sustainability, hence giving it a competitive advantage internationally' (BSI 2017: 2). Statements such as this minimize some of the agentic potential of grounded materialities in favour of transactional benefits and indicate some of the politics behind developing and implementing standards of this nature. The prioritized use of the economic repertoire fits into the operations of an organization with clear business priorities. Even the more ecocentric of the repertoires – the ethical as well as the practical – work here to enhance competitive development. Thus, responsibility and working as a pioneer in sustainability are positioned not so much about developing a more accountable way of dealing with the environment but about staying ahead of legislation and gaining a competitive advantage. Accordingly, BS 8909 is a way to 'help you run your business in a more sustainable way… that enables you to take account of the social and economic impacts your business has, as well as the more immediately obvious environmental impacts' (ibid.: 4). Even here, the balances between the repertoires view the ethical dimensions of sustainability actions as more of an afterthought.

The structures of the ordering discourses used by BS 8909 to contribute to both the UK sustainability network and to construct its own particular network are thus resolutely anthropocentric. This is especially the case as it shrouds its pragmatic repertoires in clichéd rhetoric: 'Nor is [the Standard] a pass or fail calculator; it's a guide for a journey' (BSI 2017). The framing of the discussion as a journey indicates a linear narrative process that anthropomorphizes the complex roles of agentic grounded materialities in the network. It limits the potential they have for participation and prioritizes conventions and logic that befits a resolutely anthropocentric stance. In addition, another tactic sharing many key similarities with this twisting of the practical repertoire is the use of populist frames such as celebrity endorsements: 'sustainable film attracts better talent, bigger audiences and has lower costs. It's no surprise that the film industry is fast catching up with famous actors who have been espousing environmental responsibility for years' (ibid.).

The use of these repertoires is intriguing in the context of a UK institution of standards as celebrity environmentalism, as Vaughan has argued, tends to be very much part of the greenwashing agenda of the hyper-consumptive film industry. Here, they veer any agentic potential these materialities may have in the direction of explicitly anthropocentric logic. This is even more so the case when discussion turns to economics: 'There's also money to be saved from better resource and energy use.' The business incentives environmental strategies provide reduced 'reputational risk and enhance competitiveness.

There is also a clear correlation between sustainability performance and profitability' (ibid.). This positions the adoption of sustainable strategies as a matter of economic common sense and as part of the competitive core of the industry. Finally, the discussion even includes a reference to 'the well-respected Morgan Stanley Capital World Index' (ibid.) to argue that such ideas will make financial sense for any media organization.

Quantitative readings

The discussion to date has focused on qualitative readings of promotional online material. All of the material addressed above – from Greening Film to the BBC's sustainability statements – are laden with the problematic repurposing of the repertoires for what amounts to, arguably, PR benefits. While promotional material tends to be comparatively to the point and open about its function, regulatory statements must, at least in principle, appear neutral in outlining organizational strategies. It is to this material I now turn. I will start this discussion by conducting quantitative analysis of the frequency of key words in policy documents by the main media organizations in the UK. The documents here consist of a range of corporate statements and annual reviews that outline key strategic motivations of each company. Quantitative data from these reports reveal patterns of rhetoric that can tell us more about the ordering discourses used by the organizations and whether they rely on grounded or transactional materialities.

The repertoires that construct these discourses will be used to trace power relations and balances in and outside the networks, especially in the ways material considerations integrate with managerial decisions. The repetition of certain concepts and terms will allow us to establish the KPIs of each organization to unravel the ways these nodes respond to industry developments and regulatory provocations in the wider sustainability network. In reporting keyword frequencies, I provide the top 25 terms of each document. I will edit out words that have no significant relevance for the repertoires established above. These include terms like conjunctives or words that are too abstract in their significance for the constitution of the network (work, use, year, www). Priority areas will be names of organizations, as these indicate nodal connections, and the use of terms such as business or sustainability. These rankings will generate data used to establish the general parameters for agenda setting in relation to eco- and anthropocentric priorities. Subsequently, this will be complemented with qualitative analysis of the documents.

Green initiatives in the arts

When beginning to assess the operations of the UK sustainability management network, we must start out from understanding the influence of larger external networks that facilitate the activities of the media sector. This is especially necessary as public funding for the media in the UK is often a patchwork of arts and cultural funding. Simultaneously, sustainability networks are always implicated in other regulatory or organizational networks that shape their particular directions or affect their KPIs. Thus, I will focus on the bigger picture of the arts sector initially to outline the key policy priorities and strategies for this area of the cultural industries. The analysis will reveal some of the more comprehensive sustainable management patterns that pertain to the arts as a whole. They also provide a comparative basis from where to approach the specific mechanisms developed to integrate sustainable strategies into the media industry.

We start out with *Sustaining Great Art* (2015), a document produced by Julie's Bicycle, a global environmental charity, outlining a project introducing sustainability into the arts sector between 2012–2015. Julie's Bicycle was invited by the Arts Council England, the leading arts body in the UK, to provide technical tools and training support for the project. The Arts Council has paved the way for establishing sustainability as an industrial KPI as one of its funding mandates is that organizations receiving support have to report on energy and water usage. This is a major development in the field as most of the environmental incentives to date have been based on voluntary motions. Requiring organizations to provide an environmental policy and action plan and report on their footprint ensures that the Arts Council emerges as a key OPP for the UK, at least for the arts sustainability network. For this particular venture, the data for the 2012–2015 document came from 7001 UK revenue-funded organizations, including theatres, galleries, museums, tours, festivals and concert halls. Even though the media are largely excluded here, these incentives are significant for establishing the general parameters for a sustainable media network as they chronicle the material expenditures and mitigation strategies deployed by these closely related organizations. Finally, *Sustaining Great Art* is significant as it identifies both an increasing volume of calls for environmental awareness in the industry as well as pressure to adopt environmental production measures to meet a 'cultural shift' (Julie's Bicycle 2015: 30). This is an intriguing point as it indicates an increasingly normalized presence of environmental sustainability in this sector.

The other document analysed here comes from the Green Arts Initiative (2015) (*Report 2015*), a community of organizations focused on environmental sustainability in the arts. This venture is run by Creative Carbon Scotland with Festivals Edinburgh and enhances the sharing of knowledge, ideas and experiences, and general sustainability competencies of arts organizations. The organization has over 130 members working on relevant areas including carbon emissions reduction and the establishment of Green Teams. The document is mostly focused on presenting achievements and indicators of the future plans of the organization in sustainability and arts management. As it also includes key media sectors such as film, it provides a useful contextual comparison for our discussion of media organizations in the UK as well as of the strategies of the Arts Council. Table 4.1 presents the

Table 4.1 Key words for *Sustaining Great Art* (Julie's Bicycle 2015) and *The Green Arts Initiative Report* (2015)

Sustaining Great Art	Number of occurrences	The Green Arts Initiative	Number of occurrences
Word		Word	
environmental	144	arts	34
arts	127	green	20
organisations	97	theatre	20
reporting	82	festival	18
Julie's bicycle	81	initiative	10
energy	75	art	9
sustaining	54	creative	8
creative	48	sustainability	8
sustainability	48	centre	7
art	48	community	7
council	48	environmental	7
cultural	46	carbon	6
water	42	international	6
report	41	members	6
action	39	museum	6
data	38	music	6
theatre	38	staff	6
carbon	35	company	5
resources	35	film	5
results	34	measuring	5
building	34	energy	4
co2	33	annual	4
sustainable	33	organisations	4
creativity	31	festivals	4
emissions	28	gallery	4

25 top key words from these documents. The words will be roughly grouped into grounded and transactional categories based on how they incorporate agentic material and human agency.

As these documents are targeted at enhancing industry sustainability, the word allocations reflect the particular eco-philosophical stance of the organizations. Through this, they emphasize the balance they strike between addressing environmental responsibility and organizational management. *Sustaining Great Art* has a high concentration of key words concerned with an environmentalist agenda: environmental (144, 1.72 per cent), sustaining

(54, 0.64 per cent), sustainability (48, 0.57 per cent), and sustainable (33, 0.39 per cent). Many terms that could be argued to be anthropocentric in orientation take on a more environmentalist connotation as part of their context of utterance and publication. These include terms denoting uses of materials: energy (75, 0.89 per cent), water (42, 0.50 per cent), carbon (35, 0.42 per cent), resources (33, 0.42 per cent), building (34, 0.41 per cent), CO_2 (34, 0.41 per cent), and emissions (28, 0.33 per cent). If we add up these environmental indicators, we arrive at a total of 6.7 per cent out of the total word count.

Organizational management and legislation comprise a significant part of the document as well: organisations (97, 1.16 per cent), reporting (82, 0,98 per cent), Julie's Bicycle (81, 0.97 per cent), council (48, 0.57 per cent), report (41, 0.49 per cent), data (38, 0.45 per cent) and results (34, 0.41 per cent). These are all terms that relate to the management of organizations and thus imply the transactional use of grounded materialities. Practice can also be considered as part of the managerial agenda due to its focus on anthropocentric coordination of materialities. Thus, key words such as art (127, 1.51 per cent), creative (48, 0.57 per cent), cultural (4, 0.55 per cent), action (39, 0.47 per cent), theatre (38, 0.45 per cent), and creativity (31, 0.37 per cent) indicate the necessity to meet core competencies in ways that, in this particular context, emphasize how KPIs of the arts sector can be achieved in tandem with environmental sustainability. This transactional content amounts to 9.52 per cent. From these comparative figures, we can see that the transactional paradigm is 2.82 per cent more prominent than the grounded materialities in this document.

The Green Arts Initiative has a similar allocation of focus. The following terms, which make up the majority of the document, fit with a managerial, or a transactional agenda: arts (34, 3.4%), theatre (20, 2.14%), festival (18, 1.93%), initiative (10, 1.07%), art (9, 0.96%), creative (8, 0.86%), centre (7, 0.75%), community (7, 0.75%), international (6, 0.64%), members (6, 0.64%), music (6, 0.64%), staff (6, 0.64%), company (5, 0.54%), film (5, 0.54%), measuring (5, 0.54%), annual (4, 0.43%), organisations (4. 0.43%), festivals (4, 0,43%), gallery (4, 0.43%).

The total for the transactional areas amounts to 17.7 per cent. This is a sizable percentage of the whole document and indicates the extent to which even these sustainability documents are ultimately aimed at managerial anthropocentric practice. The contrast is even more evident when compared with the grounded terms totalling 4.83 per cent of the document's key words: green (20, 2.14 per cent), sustainability (8, 0.87 per cent), environmental (7, 0.75 per cent), carbon (6, 0.64 per cent), travel (4, 0.43 per cent). The key word allocations provide an impression of the reality these policy documents consolidate.

Yet, it has to be remembered that these figures are only indicative of the rhetorical patterns of the documents and not absolute data to reveal the ideological predispositions of the organizations. Instead, they are best considered approximations that indicate the key directions of the ordering discourses in the networks these organizations generate. Thus, it will be necessary to complement this analysis with an overview of the key rhetorical content of the documents as this will also provide a clearer picture of the general context in which these particular sustainability management networks operate. I will only focus on *Sustaining Great Art* here (Julie's Bicycle 2015), largely to do with limitations of space, but also as it provides a comprehensive account of how the Arts sector approaches environmental sustainability.

While the key word analysis indicates strong emphasis on transactional ordering discourses, qualitative analysis of framing and agenda setting paints a more complex picture. *Sustaining Great Art* opens with the CEO of the Arts Council, Darren Henley, explaining the positive response the initiative has met from the industry. The emphasis is on the notion that this is the first arts funding body in the world to embed environmental sustainability into all its major funding agreements. The mandatory requirements introduced by the Arts Council and Creative Scotland have resulted in rising participation among the regularly funded organizations, which have increased in total from 14 per cent to 98 per cent between 2012–2015. As a consequence, the legislative and economic repertoires dominate *Sustaining Great Art*. While conventional motivational rhetoric ('when it comes to the environmental agenda, the arts can do it all'; ibid.: 3) is prevalent, other parts are much more directly focused on the business principles of the industry. The economic repertoire is particularly prominent throughout the document, especially in a section called 'Increasing efficiency and financial savings'. Simultaneously, these arguments are tempered with more ethical content including the Foreword, which argues that the arts are not only to 'pursue short-term gain at the expense of future generations' (ibid.: 2), but even the pragmatic repertoire is replete with discussion on economic savings: a focus on the environment is 'no longer just "nice to have" but an issue that is critical to good business' (ibid.: 3).

To support this argument, the document emphasizes the benefits gained from adopting these activities: '51 per cent of the organizations taking part reported financial benefits and 43 per cent reported reputational benefits. 70 per cent found their environmental policy useful when applying for funding and 69 per cent when engaging with stakeholders' (ibid.: 7). In 'Making the business case' the document goes to extensive lengths to argue that benefits beyond the ethical aspirations on CO_2 emissions reductions are largely financial: 55 per cent had seen financial benefits as a result of action; 67 per cent reported benefits to team morale; 40 per cent had experienced benefits to their profile and reputation; 43 per cent reported reputational benefits. This emphasis on the economic repertoire, even in the guise of reputation management, is captured by the suggestion, 'The more engaged an organization, the greater the benefits' (ibid.: 28), suggesting not so much engagement with the environment but with business management.

All this rhetoric indicates a trajectory where green measures are incorporated into an art organization's economic and reputational performance as transactional benefits. Consequently, the use of these figures as part of the transactional ordering of the network comprises instances of reterritorialization, as seen in statements such as this: 'Energy use is the biggest source of emissions, generating 97% of footprint. Over two years, reductions across the reporting group saved 12,673 tonnes of CO_2 emissions – equivalent to cost savings of £2.29 million' (ibid.: 22). The flow of the argument literally translates grounded materialities into transactional ones by positioning them as financial benefits. This flow of argumentation – the essence of the transactional discourse – also applies to the ways developments in environmental regulation are conveyed:

> [T]he findings of the latest IPCC Fifth Assessment Report are widely accepted and political rhetoric is more positive. Clean technologies and markets have rapidly expanded and global investment in clean energy increased from 268 billion dollars to 310 billion between 2013 and 2014.
>
> (ibid.: 10)

Legislation, a possibly ecocentric form of activity, in this instance leads to a set of economic benefits, resulting in a networking process that is narrated through weighty discursive sentences highlighting anthropocentric priorities.

Reterritorialization emerges as a key tactic in this document as materialities are consistently constrained by discursive acts of narration that position them as the basis for business investment instead of considering them as a value in their own right: 'The weight of material gathered over the last three years can prove that the business case and the creative case for cultural engagement with sustainability work best when they work together' (ibid.: 9). However, this sort of collaboration is never a value-free proposition as balances of power in the organizational network reterritorialize the ecocentric power of agentic materialities. All these rhetorical efforts frame sustainability initiatives as a potential problem to be overcome through economic or practical incentives. These initiatives thus hold a subservient role in the organizational hierarchy which harks back to the problematic associations created by the concept of sustainable development. Maxwell and Miller argue that 'the concept of sustainability is still commonly deployed to signify an uneasy and frankly irresponsible balance between socioeconomic development and environmental protection' (2012: 175). The contradictions inherent in the concept of sustainable development emerge at the point at which quantitative economic development overtakes environmental concerns, under which we could categorize some of the rhetoric used by the Arts Council. And as we have already suggested, in its weakest form, sustainable development becomes 'little more than "sustainable" capitalism' (Pepper 1993: 451), an argument that attests to the necessity of exploring the rhetoric used by these environmental management documents.

Even as we can identify some of the arguments used by the Arts Council and the Green Arts Initiative as part of a 'sustainably capitalist' industry, the reality constructed by their policies combine both economic and environmental considerations. For example, the data collection tools by Julie's Bicycle are a means for media organizations to 'track their environmental impacts from energy and water use and organise data analysis of carbon footprints' (Julie's Bicycle 2015: 4). Material concerns provide the basis for this tracking activity and feed into the assessment mechanisms of the industry, contributing to a distinct environmentalist agenda by the companies. They comprise the 'single biggest environmental dataset for the arts in the world' (ibid.: 10), a process that clearly highlights the knowledge-sharing benefits to be gained from studying grounded materialities. While this is inherently concerned with sharing knowledge, the purpose, at least on the page, is not about transactionality. The stated aim is to act as a distribution hub for facilitating the creation of 'carbon technologies to reduce energy consumption and CO_2 emissions, and generating more environmental awareness with staff, to inspire ideas, connect communities and other stakeholders' (ibid.: 17). These actions have not been developed due to political imposition or strategy dictated by the government, which would apply in the case of many other sectors. Instead, they can be considered as a response to the lack of financial and policy incentives for the environment by the Cameron government and subsequent caretaker administrations. According to the Arts Council, the responsibility of ensuring the principles of the COP21 are maintained now falls on 'civil society' (ibid.: 11). Cultural industries are here positioned as key participants in such a society and it is their duty to respond in kind.

To achieve this in practice requires careful targeting of key positions in the organizational management network. *Sustaining Great Art* emphasizes the virtues of leadership, both within the industry as well as inside the organization, yet, 'much of the progress has come from the middle rather than the top of organizations' (ibid.: 8) – that is, from those tasked with measuring and managing impacts rather than from executives. This has significant repercussions as it emphasizes the limitations of targeting senior management. The act of attempting to entice newcomers into the network with rhetoric drawing on the economic or the legal repertoires, for example, would not make much sense in this equation as those motivated to take environmental activities on board are most likely drawn into the network by ethical concerns. The suggestion in these documents is that sustainable action is often a result of motivated individuals and top-down management only works at a certain level, i.e. when economic or reputational costs are on the line.

Yet, a key argument made by the Arts Council is that sustainability needs long-term funding and various forms of governmental and institutional support to succeed. To date, much of the activity has been premised on voluntary mechanisms and ethical participation as there simply is no real urgency to do so in terms of infrastructural guidance. Frequently, however, this role falls to cultural organizations as fluctuations in domestic politics and the persistent uncertainty over commitments to renewables and other green incentives results in an emphasis to 'foster confident decision-making that looks beyond political and funding cycles' (ibid.: 35). Thus, the often contradictory messages presented in the arts and culture sector indicates a reality that struggles to find a sense of balance over the most appropriate forms of strategy for incorporating environmental sustainability into production strategy. Transactional materialities receive more coverage over grounded ones when inspected through the key words, which indicates the role of these documents as business strategies. Yet, qualitative focus shows both authentic concern for the environment and pragmatic necessity to frame the debates in business rhetoric. This is, of course, the reality in which they must operate, where CSR is always and invariably a business consideration, as much as it is an ethical concern.

Bibliography

BAFTA. 2015. *We Know How to Make Greener Telly: Now Let's Get Down to Business*. London: BAFTA.

BAFTA. 2017. *The Bafta Albert Consortium*. Available at: http://wearealbert.org/about/the-consortium (accessed 11 November 2017).

BBC. 2017. *Environmental Sustainability*. Available at: www.bbc.co.uk/responsibility/environment (accessed 11 November 2017).

Bottrill, C., Liverman, D. and Boykoff, M. 2010. 'Carbon soundings: greenhouse gas emissions of the UK music industry', *Environmental Research Letters*, 5: 1–8.

BFI. 2009. *Film Forever 2010–2015*. London: BFI.

BSI. 2017. *Sustainable Film*. Available at: https://shop.bsigroup.com/Browse-By-Subject/Environmental-Management-and-Sustainability/Sustainability/Sustainable-film-with-BS-8909/ (accessed 11 November 2017).

Caldwell, John. 2013. 'Para-industry: researching Hollywood's Blackwaters'. *Cinema Journal*, 52(3), 157–165.

Caldwell, John. 2017. 'Spec world, craft world, brand world'. In Curtin, Michael and Sanson, Kevin (eds) *Precarious Creativity: Global Media, Local Labour*. Los Angeles: University of California Press, pp. 33–49.

Entman, Robert. 1993. 'Framing: toward clarification of a fractured paradigm', *Journal of Communication*, 43(4): 51–59.

Fuller, Matthew. 2005. *Media Ecologies: Materialist Energies in Art and Technoculture*. Cambridge, MA: The MIT Press.

Genette, Gerard. 1997. *Paratexts*. Cambridge: Cambridge University Press.

Grainge, Paul. (ed.) 2009. *Ephemeral Media: Transitory Screen Culture from Television to YouTube*. London: Routledge.

Green Arts Initiative. 2015. *Report 2015*. Edinburgh: The Green Arts Initiative.

Greening Film. 2011. *Greening the Screen: How to Go Green without Going into the Red*. London: Greening Film.

Greening Film. 2017. 'The Legal Driver'. Available at: www.greeningfilm.com/ (accessed 11 November 2017).

Grey, Jonathan. 2010. *Show Sold Separately: Promos, Spoilers and Other Paratexts*. New York: New York University Press.

Hey, Christopher. 2009. 'EU environmental policies: a short history of the policy strategies', in European Environmental Bureau (ed.) *EU Environmental Policy Handbook*. Brussels: European Commission, pp. 17–31.

Hills, Matt. 2015. *Doctor Who: The Unfolding Event*. Basingstoke: Palgrave.

IPPR. 2007. *Warm Words: How Are We Telling the Climate Story and Can We Tell It Better?* London: IPPR.

IPPR. 2011. *Warm Words II: How the Climate Story Is Evolving*. London: IPPR.

Julie's Bicycle. 2015. *Sustaining Great Art: Environmental Report*. London: Julie's Bicycle.

Law, John. 2007. 'Actor network theory and material semiotics'. Available at: www. heterogeneities.net/publications/Law2007ANTandMaterialSemiotics.pdf (accessed 11 November 2017).

Lipschutz, R. and Rowe, J. K. 2005. *Globalization, Governmentality and Global Politics: Regulation for the Rest of Us?* New York: Routledge.

Maxwell, Robert and Miller, Toby. 2012. *Greening the Media*. Oxford: Oxford University Press.

McCombs, Maxwell, Shaw, Donald, and Weaver, David. 1997. *Communication and Democracy: Exploring the Intellectual Frontiers in Agenda-Setting Theory*. New York: Lawrence Erlbaum.

Media CSR Forum. 2013. *Does It Matter: Material, Strategic or Operational?* London: Carnstone.

Pepper, David. 1993. *Eco-Socialism: From Deep Ecology to Social Justice*. New York: Routledge.

Walker, Janet and Starosielski, Nicole. 2016. *Sustainable Media*. New York: Routledge.

5 The sustainability rhetoric of film and television organizations

Introduction

The arts organizations explored in Chapter 4 highlight many of the key repertoires and ideological preoccupations that shape sustainability management as a compromise between eco- and anthropocentric approaches to environmental management. These dialogues continue in the film and television industry which also struggles with finding its particular approach to sustainable capitalism. The British Film Institute (BFI) is the leading body for film in the UK and an important central node in this network. It provides funds for the majority of domestic productions as well as different types of support for the exhibition and distribution of films. The BFI has taken the lead in integrating sustainability into the film sector and works in collaboration with other key agencies. These include organizations such as Film London and Creative Skillset, which together have formed Green Screen, a consultancy that provides the industry with advice on sustainability. Most of these efforts are coordinated in practice by a consulting organization called Green Shoot, which provides both training and calculators for film shoots. They in turn collaborate with the EU-wide Cine-regio network, a leading body for sustainability in European cinema composed of 45 film institutes. The work of Green Shoot, Cine-regio, etc. will be addressed in more depth in Chapter 9 on film and television, but for now, we will focus on charting some of the key features of the UK's film and television network. We do this by starting out with organizational policy documents that provide a glimpse into the top-down management of environmental sustainability in the film and television industry. We then move on to more practice-based industry manuals that present a much more grounded understanding of environmental production practices. This part of the discussion concludes with a brief case study of the BAFTA carbon literacy workshop, which provides perhaps the most incisive picture on how to integrate environmental policy with production considerations. The chapter concludes with an exploration of managerial documents from CSR consultancies that provide a more emphatically management-focused approach.

The BFI

Trialling the BS 8909 (BFI 2015) is an appropriate starting point for this ana-lysis as it is a key document in instigating and consolidating the UK envir-onmental management network. Not only is it a document outlining developments in generating an environmental sustainability strategy for the film sector, but it is one of the most comprehensive overviews of these developments in the sector. The document, an exemplary case of an estab-lished strategy to implement sustainability into the sector, outlines the development of the BS 8909 sustainability management standard and pro-vides data on monitoring productions that have received public funding from the BFI for 2013–2014. It also provides case studies of productions where sustainability standards were integrated into production practice. These areas make it an essential component of the network.

The scope of *Trialling the BS 8909* is also intriguing as it contextualizes environmental policy with wider debates on climate change, scientific data and the Paris Climate Conference in 2015. According to the document, environmental activity is the duty of every part of the 'global industry as the ongoing unlimited burning of fossil fuels is the cause of climate change' (ibid.: 2). In doing so, it positions environmental management and labour practice as an ethical necessity the industry must be able to address. The BFI suggests that 'a co-ordinated approach to sustainability using BS 8909 will help us all meet UK carbon budgets and lead to greater efficiencies and long-term cost savings so that budgets can be used to better support the film industry' (ibid.: 2). The statement positions the document's strategy as part of a Plan-Do-Check-Act cycle that uses Management System Standards such as ISO 14001 (environmental management) and ISO 9001 (quality assurance) to measure the performance and adaptation of environmental protocols in different industrial sectors. The BS 8909 is the film industry version of this strategy and thus provides a mechanism for film produc-tions to address their footprint in ways that meets established regulations and standards.

The categorization of the key words of the document into grounded and transactional materialities (to correspond with the environmental and the managerial categories used above, respectively) reveals some of its key paradigmatic directions (Table 5.1).

Grounded key words amount to 7.55 per cent of the total document with transactional ones at 10.27 per cent. Thus, the emphasis is on transactional matters, but only by a small margin. The balance can be explained by the document's focus on addressing production staff, including directors and crew. Furthermore, a large part of the document is devoted to three film production case studies comprised of *Free Fire* (2017), *Swallows and Amazons* (2016) and *City of Tiny Lights* (2017). The emphasis on production practice results in human resource management taking up a large part of the document. While this emphasizes a transactional approach, simultaneously, this

Table 5.1 Key words for *Trialling the BS 8909*

	Key word	Number of occurrences	(%)
Transactional	production	76	2.04
	management	41	1.10
	film	34	0.91
	filming	32	0.86
	crew	29	0.78
	support	28	0.75
	programme	25	0.67
	runner	24	0.64
	productions	22	0.59
	team	21	0.56
	system	20	0.54
	report	19	0.51
	industry	17	0.45
	prep	16	0.43
Grounded	sustainability	57	1.53
	green	48	1.29
	environmental	39	1.05
	greenshoot	35	0.94
	waste	19	0.51
	set	15	0.40
	water	15	0.40
	transport	14	0.38
	carbon	13	0.35
	catering	13	0.35
	recycling	13	0.35

(BFI 2015).

practical angle means that material concerns have a much more pronounced role here than in the arts documents which were more emphatically targeted at organizational management. Consequently, discussion of production management activities, such as catering and transport, take the discussion much closer to the agentic potential of grounded materialities than transactional considerations. While management will pay considerable attention to budgeting for sustenance and costs of petrol, for example, the act of resourcing food and curbing waste are more inherently ecocentric rather than transactional. This indicates that a practical approach will inevitably be more ecocentric as it has to work with resources first-hand, which provides more agentic power to grounded materialities.

Yet, the simultaneous use of phrases like 'greater efficiencies and long-term cost savings' (ibid.: 2) as a key motto for the BFI makes it clear that this document still aims to develop production policy that needs to find a balance between the environmental and the economic. Such debates are very evident in the document's simultaneous emphasis on top management and the need to employ green runners as essential components of a sustainability management system (SMS). The rhetoric continues to balance between the eco- and the anthropocentric and makes it clear that human resource management plays a significant role in environmental sustainability. Yet, arguably, the most intriguing notion of this document is that once we move to the level of production coordination, the rhetoric shows an increase in the agentic role of grounded materialities to accommodate the level of power they hold in practical production coordination.

The BS 8909

The emphasis of *The BFI Sustainability Review* is, as expected, explicitly on production management, yet some of the balances between the eco- and the anthropocentric can also shift according to the orientation of the documents, for example, from production to policy. One such document is *The BS 8909 Case Study* (BSI 2014), produced for the British Standards Institute (BSI). The BSI is a business standards company and a member of the International Organization for Standardization (the ISO) that provides a variety of industrial sectors with internationally recognized standards of best practice such as ISO 14001. BS 8909 is the first standard for sustainable film production in the world and was commissioned by the UK Film Council in 2009. While the BFI has taken over these responsibilities from the Film Council since 2011 and run them in association with Green Shoot, the early steps of this venture were completed in collaboration with the Department of Business, Innovation and Skills, 'as part of its ongoing commitment to supporting innovation in the UK' (ibid.: 4). The BS 8909 standard was thus designed as a particularly business-focused strategy. Amanda Nevill, the Director of the British Film Institute, summarizes these imperatives:

British Standard 8909 will provide the framework for BFI to manage our environmental, social and economic impacts on an ongoing basis. BS 8909 is a cutting edge sustainability tool that the British Film industry should be proud of and embrace to maintain a healthy and competitive film industry.

(ibid.: 1)

The framing of this debate indicates the ways the BS 8909 is used to fulfil the legal, the economic and the ethical imperatives of the sector by merging it with practical necessities. This is not only about promoting beneficial environmental values but also about a healthy industry building on its existing 'sustainability credentials by adopting formally-recognized best practice' (ibid.: 2). *The BS 8909 Case Study* discusses energy reductions across the film production and exhibition experience and indicates areas where performance could be improved. It is clear that grounded materialities play a significant role in the data on which the standard is based, even though they feature less in this document. Most of it focuses on key managerial areas such as risk assessment, performance matrices, job appraisals and establishing procedures for different professional roles in the organization. The use of such managerial repertoires is important as it enforces a sense of normality – of standard practice – on sustainability.

As part of this normalizing rhetoric, Duncan McKeich, facilities manager for the BFI, suggests 'sustainability needs to be considered in a similar way to Health & Safety, that is, as basic good practice, integrated into all organizational decision making' (ibid.: 3). These perspectives complement the document's adherence to the Brundtland Commission's approach to sustainable development as a means to continue business as usual in a way that does not take more out of the planet than is required. In a somewhat similar way, Amanda Nevill suggests that the framework will help the industry meet 'environmental, social and economic impacts' (ibid.: 1) but does so in a way that does not compromise its ability to meet the industry's core competencies. Accordingly, a focus on the key words shows a distinct emphasis on management with transactional key words at 15.9 per cent and grounded at 5.8 per cent (Table 5.2).

Even with the inclusion of impacts (which could fit in the managerial categorization as well), it is clear that the focus is on the transactional benefits sustainability brings. This is not surprising considering the role of the BSI and the job it was asked to perform in developing the Standard. The equation of sustainability management on a par with health and safety is part of this transactional logic, where the professionalization of sustainable management may make it appear more anthropocentric than it actually is. In many ways, standards like the BS 8909 comprise a new code of conduct that reinforces the sense that sustainability needs to be translated into the economic and legal repertoires to make an impact. The managerial and economic repertoires give it an edge that may appear problematic from an ecocritical perspective, but these are also necessary strategies if these organizations wish to take the core competencies of the media business into account.

Sustainability in television

The BS 8909 document provides the most obvious example of a transactional ordering discourse shaping the UK sustainability network. *The BFI Sustainability Report* and *The BS 8909 Case Study* focus on senior managerial

Table 5.2 Key words for *The BS 8909 Case Study*

	Key word	Number of occurrences	(%)
Transactional	BFI	39	2.91
	BSI	19	1.42
	management	16	1.20
	standards	15	1.12
	British	12	0.90
	film	12	0.90
	industry	12	0.90
	standard	12	0.90
	organizations	10	0.75
	performance	8	0.60
	institute	7	0.52
	social	7	0.52
	system	7	0.52
	development	7	0.52
	trial	7	0.52
	director	7	0.52
	issues	6	0.45
	business	5	0.37
	committee	5	0.37
	economic	5	0.37
Total			15.8
Environmental	sustainability	43	3.21
	impacts	8	0.60
	energy	7	0.52
	environmental	7	0.52
	sustainable	7	0.52
Total			5.8

(BSI 2014).

levels even as they adopt slightly different perspectives in doing so. The former has a much more prominent focus on production management whereas the latter is inherently concerned with the business case. These approaches are replicated across the television industry. The British Academy for Television Arts (BAFTA) is leading the integration of sustainability in this part of the UK sustainability network. They have been working with the NGO Julie's Bicycle on developing the carbon calculator Albert that is now a leading international benchmark for the television industry and provides a central obligatory passage point (OPP) for the environmental media network in the UK.

I focus on the BAFTA document *Environmental Sustainability: Ready Telly Go!* (BAFTA 2015) which is a year four report compiled for the Albert Consortium. This is a particularly useful example to address at this stage as the document is advanced enough to showcase a solid evaluation of industry acceptance of environmental standards but still one that is at a relatively developing stage in implementing them. In doing so, the document outlines the 'footprint of the industry, summarises the consortium's key achievements and outlines plans to lead the industry towards a more sustainable future' (ibid.: 2). Kevin Price, the Chair of the Consortium and CEO of BAFTA, suggests that the aim of the consortium is to reorganize 'our systems to design waste and emissions out of the equation and communicate our progress effectively' (ibid.: 2). The development of best practice guidelines for the industry and making the case for their wider adoption is captured through a balanced combination of grounded and transactional materialities as reflected in the key words (transactional at 10.34 per cent, grounded at 8.95 per cent) (Table 5.3).

While areas focused on human-led management occupy a large percentage of the key words, the diversity of material considerations in both infrastructural and production planning contributes a clear sense of material agency to the rhetoric. This is also reflected in the content of *Environmental Sustainability: Ready Telly Go!* which includes in-depth information on the footprint of the television sector as a whole. These cover areas from the use of solar-powered technology in shooting *Planet Earth* to swapping the energy solutions for lighting on *Downton Abbey*. Other parts include informatics on the total footprint of producing one hour of television and visual comparisons between the annual CO_2 emissions of the industry with famous landmarks in London. These are all transactional materialities which provide translations of grounded materialities in easy-to-comprehend ways. As the document is aimed at a wide range of professionals in the television sector, it needs to achieve a balance between management and environmental considerations, and the fact that it is a year four document focusing on a relatively advanced stage of development allows it to move from business and management pitches to more concrete efforts with fluency.

Environmental Sustainability: Ready Telly Go! acts in much a similar way as *The BFI Sustainability Report* in that it caters for the sector as a whole through a relatively balanced integration of grounded materialities. As it is focused

Table 5.3 Key words for *Environmental Sustainability: Ready, Telly, Go!*

	Key words	Number of occurrences	(%)
Transactional	production	39	2.44
	industry	28	1.75
	change	20	1.25
	Albert	18	1.13
	support	10	0.63
	productions	9	0.56
	make	8	0.50
	create	7	0.44
	opportunity	7	0.44
	training	7	0.44
	consortium	6	0.38
	people	6	0.38
Total			10.34
Grounded	carbon	30	1.88
	environmental	14	0.88
	sustainability	13	0.81
	climate	13	0.81
	sustainable	12	0.75
	emissions	11	0.69
	footprint	10	0.63
	power	8	0.50
	studio	8	0.50
	green	7	0.44
	impact	7	0.44
	energy	5	0.31
	location	5	0.31
Total			8.95

(BAFTA 2015).

on both production and policy, these materialities play a more agentic role in the network than in *The BS 8909 Case Study* which is more explicitly concerned with strategic or organizational management directives. Yet, at the same time, we can ask whether these strategies are ultimately much different. To ensure appropriate uptake, the ethical and the practical repertoires are mobilized as managerial tools where the aim is to encourage 'programme makers in changing the world for the better. The industry must be informed about the challenges facing all humanity' (ibid.: 2). Arguably, the rhetoric used here suggests that BAFTA's strategic development integrates management and materiality in ways that refuses simplistic reterritorialization of material agency for transactional purposes. At the same time, the economic and the legal repertoires reinforce the sense that these strategies are rational industry goals intended to 'generate financial savings for all involved [as well as make] sustainability easy, accessible and interesting to the communities we serve'.

Yet, even as the BAFTA operations may struggle to strike an appropriate balance, they state that 'the services and practices currently available and employed by productions presently' (ibid.: 3) make achieving zero emissions unattainable with the present strategies implemented in the industry. Thus, while it is encouraging to see that the stated goal for the consortium is a resolutely material one based on cutting out emissions in total, the fact that these strategies will not be sufficient for the future raises pertinent questions. To counteract these problems, the document poses BAFTA's work on enhancing Free Carbon Literacy training and increasing the share of renewable energy in the creative industries as new strategies aimed at addressing ignored areas and normalizing sustainability as part of everyday production practice. Yet, they also testify to the early stage of development of environmental sustainability strategies in the media industry, even in these more advanced contexts.

The BBC

Environmental Sustainability: Ready, Telly, Go! and *The BFI Sustainability Report* aim to provide a combination of production and managerial responsibility, and minimize the economic and regulatory repertoires emphasized by documents like *The BS 8909 Case Study*. Even as they manage to achieve a more ecocentric orientation, this approach is complicated by the BBC, as a central organization of the UK television network, that uses both approaches. I focus on two documents, *Albert – Programmes that Do Not Cost the Earth* (BBC 2011), a guide developed by the BBC to address producers of programming, and *The Environment – A Sustainable Approach* (BBC 2015), the BBC's CSR study. *Albert – Programmes that Do Not Cost the Earth* provides a practical outline for the industry to consider sustainability in each step of the production. The key prerogative of the guide is 'to help you make your programme as sustainable as possible' (BBC 2011: 4). It is structured in the form of a model with which producers can trace emissions through the production cycle. Accordingly, it has a heavy pragmatic emphasis, covering areas where the producer may be able to integrate sustainable measures into practice. The results of the key words confirm the balance between the grounded and the transactional materialities, or the environmental and managerial approaches (transactional at 9.3 per cent, grounded at 7.51 per cent) (Table 5.4).

While there is more emphasis on transactional aspects of production management, the grounded components of the document provide considerable diversity of scope. The focus on each material component of the production process results in comparatively high frequencies for areas like energy, lighting, power and paper. When these materialities feature at such a level, it is not unfeasible to suggest that they have considerable agentic potential in shaping the directions of this network. The consequence is that the document allows managerial decisions to be coordinated by the demands of the material base. Such arguments are especially prominent in the ways practical rhetoric is used to connect the economic and the environmental in what the company calls the 'three Rs of sustainable production':

> Reduce: don't use as much of anything. Whatever it is, use less.
> Reuse: adapt what you've got rather than buying or making something new.
> Recycle: if you have to throw something away, then recycle it. Waste disposal – even recycling – is the last resort. If you've got waste which can't be recycled then dispose of it carefully.
>
> (ibid.: 4)

These Rs emphasize grounded materialities as an essential part of the tactics proposed by *Albert – Programmes that Don't Cost the Earth*. This strong ecocentric approach feeds into the more obviously managerial goal of leadership of the industry. It suggests that

> a sustainable production happens because there's commitment to a plan and everyone in the team is aware of it. It has to come from the top. Senior staff should discuss and agree on the show's sustainability goals at a very early stage'.
>
> (ibid.: 4)

This emphasis on management makes sense from the perspective of adopting environmental strategies on set where time and resources can be limited and need to be carefully coordinated. Here, it is up to the 'senior staff to take a strong lead on sustainability, sharing goals with cast and crew, measuring their carbon footprint and adopting sufficient low carbon production techniques to address the overall impact of the programme' (ibid.: 4). The emphasis on an ecocentric approach is reflected in the publication title that deprioritizes economic and legal concerns – so prevalent in the film industry – all while allowing the grounded materialities sufficient agency.

Table 5.4 Key words for *Albert – Programmes that Do Not Cost the Earth*

	Key words	Number of occurrences	(%)
Transactional	use	122	2.72
	Albert	58	1.29
	section	53	1.18
	production	38	0.85
	BAFTA	28	0.62
	checklist	23	0.51
	make	23	0.51
	check	19	0.42
	studio	19	0.42
	used	19	0.42
	cast	16	0.36
Total			9.3
Grounded	travel	30	0.67
	waste	30	0.67
	transport	28	0.62
	carbon	27	0.60
	food	23	0.51
	set	22	0.49
	sustainable	20	0.45
	equipment	19	0.42
	office	19	0.42
	lighting	18	0.40
	location	18	0.40
	air	17	0.38
	energy	17	0.38
	paper	17	0.38
	power	16	0.36
	recycling	16	0.36
Total			7.51

(BBC 2011).

Finally, the task of these documents is to convince the industry to adopt environmental sustainability as standard practice. Developing OPPs like Albert is significant, as these translational means not only translate agentic materialities into rhetoric suitable for business, they also act as standardizations of environmental accounting into production management. As adding yet another measurement tool on top of all the other measurement systems the industry uses would arguably be met with hesitation, the BBC uses a range of repertoires to make the case for adopting environmental best practice. Key frames are areas such as the ability to generate a 'positive impact' and adapting to what will be the 'new norm'. The incentive is 'to do the right thing when you're under no legal obligation to do so' (ibid.: 3). The rest of the guide argues that all information will be kept 'as practical as possible' with specific sections devoted to negotiating with the supply chain as well as the crew and staff. To emphasize the centrality of practical necessity, the guide has parts on office management, printing, consumables and travel. But all this comes with a caveat: 'Sometimes doing the right thing will cost money. Other measures may save you a little bit' (ibid.: 3). The ethical, the economic and the practical coalesce here while the legal is avoided as there are no real pressures on the industry from this adoptive standpoint.

The legal imperative has become standard practice in 2017 for all production companies, both internal and external, working with the BBC, following the inauguration of mandatory reporting for the Arts Council England and Creative Scotland. The 2014 document *Albert – Programmes that Don't Cost the Earth* provides an impression of the networking activities that were required as a basic framework for this policy from a time when such measures were not fully implemented, and when they were still considered insignificant compared to the business priorities of media organizations. As the normative expectations at the time were to prepare management to anticipate resistance from the industry, this established the industrial reality of environmental sustainability as being about obstacles to developing policy. Sustainable policy was thus about convincing stakeholders of the use value of these actions for the television industry, where the combination of ethical and financial rhetoric reflected the general balance these documents would try to attain. As we will see, such actions only tend to work to an extent, and even if the network had by this time become consolidated, its constitution was still fragile. We will suggest that more hands-on regulation would be the most productive industry approach and return to these questions in later sections, but for now, we focus on other, more regulatory, means through which the television industry has tried to integrate sustainability into its infrastructure.

The BBC and CSR

In contrast to these production guides and especially the approach of *Albert – Programmes that Don't Cost the Earth*, the BBC's *The Environment: A Sustainable Approach* provides a CSR overview of the company's work through a more managerial perspective, befitting as it was produced by the BBC's Outreach and Corporate Responsibility team. Yet, despite the limitations of the managerial approach with *The BS 89 8909 Case Study*, the environment is clearly a priority here – potentially reflecting its role as a *promotional* corporate policy document for the BBC. Accordingly, the document discusses areas such as the 'resources we use and the CO_2 emissions associated with our energy consumption, the travel we undertake and the waste we generate' (BBC 2015: 2). This is an encouraging sign that it does not consider the environment as a subsection of a larger focus on sustainability, as many similar organizations do with their sustainability strategies which mostly subsume the environmental into social and economic concerns.

The aim of the document, in contrast to the producer- and management-led considerations of *Albert – Programmes that Do Not Cost the Earth*, is to educate the public. This fits with the BBC's public service mandate concerning 'how we run our business responsibly and how this work helps us to meet our six Public Purposes as set out by the Royal Charter and Agreement – the constitutional basis for the BBC as presented to Parliament' (ibid.: 2). The document includes short practical case studies for Albert involving *Casualty* (the first programme to gain a three-star rating for implementing sustainability across all of the production) and *Springwatch*, both of which have used extensive carbon reduction measures in their production. In addition, it covers shows such as *The Great Big Energy Saving Challenge* and *Costing the Earth* that emphasize the communicative responsibilities of the organization in meeting its aspirations to be a 'steward of the environment'.

The BBC has collaborated with BAFTA on Albert and much of the document focuses on this mission, encapsulated by the slogan 'Our industry, our work'. The slogan sets the agenda as one focused on enhancing knowledge of sustainable production practice in the industry. Coverage of the Media Greenhouse venture provides a para-industrial justification for the document, linking the BBC not only with BAFTA but other key broadcasters Channel 4, ITV and Sky. These collaborative projects are especially significant for the television sector as they motivate the industry to pool together around the theme of environmental sustainability. The collaboration provides the Media Greenhouse with data from a range of operations and gives 'enormous insight into the impacts of the production process' (ibid.: 2). The data are used to develop Albert+ (a more comprehensive version of the original calculator system), where the motivation for gaining this 'gold star' is in the reputational benefits it can generate as well as a 'celebration' of achievements of producers in embracing the sustainability challenge (ibid.: 2). Thus, the document acts as a form of reputation management that caters for more senior-level personnel at production organizations as well as to the external organizations that require both transparency and accountability from the BBC.

Another section of the document focuses on 'Our staff, our business' which explores ways for motivating staff through the HR management strategy 'The Difference' and through a film called *We Can't Make a Difference Without You*. These sections focus on areas like the BBC Worldwide service, ensuring that sustainable practice becomes 'business as usual' (ibid.: 21). Other parts talk about the BBC Academy and the Sustainable Production team and the sustainability training they have implemented across the organization. As the focus of *The Environment: A Sustainable Approach* is on communicating its responsibilities to the public and to its various stakeholders, the emphasis here is on transactionality. Despite its inclusion of case studies and short examples of agentic grounded materialities in production practice, the key word allocations of the document reflect some of these overtones (transactional at 12.29 per cent, grounded at 3.97 per cent) (Table 5.5).

Despite the lack of space afforded to the material base of production in this document, the managerial impetus should not be disregarded out of hand as it contributes a central mechanism for instigating and organizing these activities. It also works as a key means of oversight over labour and procurement considerations and thus acts as an essential part of an operational network. The role of grounded materialities in these organizational relations documents (both managerial and external facing) tends to be relegated to secondary status in favour of managerial priorities. This should not be taken as anything but a reflection of these documents as ordering discourses trying to coordinate the network. The use of legal, financial, ethical and practical repertoires modulates the necessity of environmental sustainability for particular stakeholder groups, which suggests that sustainability is always predicated on a particular version of reality created for specific target demographics. This reality can range from operational strategy governing the operations of BSI to a production strategy for the BFI, from industrial oversight at BAFTA to the responsibilities of a public organization at the BBC. Ultimately, this confluence of perspectives provides an impression of the scope of ordering discourses used to consolidate the UK network. While the film network has made valid efforts to integrate sustainability into production practice, their gains with carbon calculators and industry leadership rely much more on business logic than is the case with the BBC and BAFTA – especially as the latter have started to emphasize training initiatives and other environmental literacy measures.

The different types of material expenditures for film and television – resources used and emissions generated – will be addressed in later chapters. For now, it is most significant that the environmental sustainability strategies of the film and television industries follow similar pathways in conceptualizing the agentic roles of grounded and transactional materialities. As we have seen, these are predicated on a struggle for balance that needs to leave room for both ethical aspirations and industrial realities. In facilitating connections between different nodes of the network, these strategic documents act as top-down management discourses organizing the UK sustainability network for film and television. The key areas for agenda setting based on these documents consist of sustainable development, leadership

Table 5.5 Key words for *The Environment: A Sustainable Approach*

	Key word	Number of occurrences	(%)
Transactional	BBC	88	2.73
	production	54	1.67
	work	27	0.84
	sustainability	25	0.77
	environment	22	0.68
	programmes	19	0.59
	staff	19	0.59
	industry	17	0.53
	programme	16	0.50
	use	16	0.50
	using	15	0.46
	impact	14	0.43
	responsibility	14	0.43
	working	14	0.43
	audiences	13	0.40
	business	13	0.40
	team	11	0.34
Total			12.29
Grounded	energy	31	0.96
	Albert	29	0.90
	sustainable	28	0.87
	carbon	18	0.56
	environmental	17	0.53
	power	14	0.43
	footprint	11	0.34
	set	11	0.34
Total			3.97

(BBC 2015).

and business, with ethics and environment featuring often in rhetoric but less in volume. The emphasis on management, perhaps understandably, has resulted in many of these policy documents framing sustainability as another form of organizational management, resulting in them providing a more nuanced and business-oriented version of the rhetoric seen in the promotional material.

Industry manuals

Clearly then, these policy documents adopt slightly different angles on the balance between the eco- and the anthropocentric, depending on their purpose and target demographics. Yet, how does this balance appear in documents focusing on practice? To address this area, the analysis will now move to explore industry manuals designed to implement green practice on set. These documents focus mostly on covering the experiences of producers and line management in adapting green protocols into their work practice. In doing this they are not as explicitly concerned with policy as the top-down documents and adopt a more consciously creative take on sustainable production practice. They provide an alternative view of the sustainability network by focusing on the 'selling' of sustainable production strategies to different parts of the production team.

For example, the use of repertoires appears different from the top-down management and reveals some of the dynamics of agency on this more grounded level. To illustrate, the *Inspiration and Information for a Sustainable Screen Arts Industry* (Greening the Screen 2015) is part of the Greening the Screen Event follow-up pack provided by BAFTA and the BFI. While it opens with Carol Comley, the BFI Head of Film Policy, explaining the 'growing passion' (ibid.: 2) for sustainability, the majority of the document contains case studies where producers and crew, the individuals in charge of procurement and other on-set management, share their experiences of sustainable practice. Thus, the document provides hands-on guidance on 'how to green your production' (ibid.: 7) as well as sourcing a green value chain. It concretely addresses the pragmatic circumstances of its audience by suggesting that

> [It] is difficult for the budget holder to spend more money to reduce their impact. Choosing not to act sustainably on these grounds, however, is convenient, easy and often done. We need to do vastly more, we have our heads in the sand.
>
> (ibid.: 4)

Thus, economic logic and ethical rhetoric combine to provide 'content makers and cultural leaders' (ibid.: 5) a means to transition to a sustainable low carbon future. The key words for the document highlight its difference from both policy and promotional material (transactional at 9.05 per cent, grounded at 11.29 per cent) (Table 5.6).

Table 5.6 Key words for *Inspiration and Information for a Sustainable Screen Arts Industry*

	Key words	Number of occurrences	(%)
Transactional	production	14	1.44
	crew	8	0.82
	industry	13	1.34
	local	7	0.72
	producer	7	0.72
	productions	7	0.72
	support	7	0.72
	practice	6	0.62
	screen	6	0.62
	production	5	0.51
	act	4	0.41
	information	4	0.41
Total			9.05
Grounded	carbon	16	1.64
	sustainable	16	1.64
	sustainability	14	1.44
	green	12	1.23
	power	8	0.82
	footprint	7	0.72
	lighting	7	0.72
	energy	6	0.62
	set	6	0.62
	recycling	5	0.51
	transport		0.51
	Albert	4	0.41
	environmental	4	0.41
Total			11.29

(Greening the Screen 2015).

In comparison to the organizational policy documents, there is a clear focus on environmental/material concerns in this guide, reflecting the document's focus on practical considerations in day-to-day production management. The framing of these concerns is less to do with managerial rhetoric and more focused on practicalities as the closer we get to tangible production arrangements, the more the grounded materialities will influence the rhetoric. This is also evident in incorporating practical suggestions from the crew into policy developments, as the framing of environmental sustainability is no longer presented as a KPI in terms of a company's reputation and profile but becomes a factor to do with evaluating the costs of environmental decisions on production management and budgeting. The use of instructional direct address ('be aware of any additional costs, they are manageable with advanced planning') frames these practices as potential financial costs but they are also positioned as pragmatic and sensible decisions ('setting sustainability criteria as a key deliverable is a quick "win"') (ibid.: 7). The process here certainly includes the incorporation of materialities into managerial rhetoric, but the emphasis is resolutely on how to deal with the challenges that the grounded materialities pose. On this level of the media management network, the practical repertoire has the largest impact on network consolidation as it is mobilized to incorporate the demands and challenges of these agentic materialities on production management.

The role of agentic materialities changes as we go through the different layers of the network. While the environmental and the managerial do not exist in an adversarial relationship but often complement one another, the more focused the documents become on practice, the higher the concentration on grounded materialities becomes. This is not a surprise, of course, as investment in material processes increases the closer one works with them. Accordingly, in *Inspiration and Information for a Sustainable Screen Arts Industry*, the practical repertoire dominates the discussion to reflect the necessity of accounting for material implications outside of their role in a production meeting organizational KPIs. While costs will of course matter, the main focus is on meeting ethical standards and, thus, the particularities of these materialities arguably shape the behaviour of production managers, and consequently of this area of the network.

These patterns continue in other industry manuals. *Factsheet 10: Sustainable Power for the Screen Arts on Location* (Media Greenhouse 2014), by Media Greenhouse and Powerful Thinking, provides a to-do guide on how to make 'positive choices' on location uses of power and resources. The *Factsheet* provides insights into educating the crew on, for example, the most appropriate sources of energy and discusses the sourcing of biodiesel from waste vegetable oil and the benefits of pedal power, among other relevant topics. The document is pitched at the producer level with frequent use of direct address through phrases like 'your production' (ibid.: 1). Most of the rhetoric is more precisely aimed at line managers or production specialists with considerable level of know-how of in-depth production processes. There are no mentions of costs and the document suggests the best way to curtail power consumption is to 'produce the most detailed power specification you can in advance' (ibid.: 1).

This approach produces a network that is much more specific in scope than those designed to facilitate organizational management, described above. It emphasizes collaboration between producers in charge of budgeting and line managers in charge of daily production arrangements. Here, economic costs of budgeting for the complexity of green shoots is sidelined in favour of ensuring that the specificities of each type of grounded materiality is accounted for. The mode of address links producers with other production staff by asking them to exchange in-depth data on material expenditure that ensure everyone is responsible for their department's operational consumption: 'you must ask your crew on how much power they will draw' (ibid.: 1). Through these instructions, the document goes to great lengths to explain each energy source and the environmental benefits of, for example, waste vegetable oils. This discussion is often focused on efficiency measurements or specific requirements of a shoot (i.e. portable hydrogen fuel cells or photovoltaic cells for solar power, and how these would be implemented on mobile sets). The instructions explain how specific fuel cells (such as hydrogen ones) would provide a competitive advantage over generators on wildlife shoots, for example, whereas biodiesel generators would be best for productions seeking to improve their environmental performance without altering production protocols (for example, on studio-based shooting). As we can see from the key words, material considerations are at the forefront here (transactional at 4.04 per cent, grounded at 18.16 per cent) (Table 5.7).

While grounded materialities dominate the discussion, different repertoires are used to mediate their agentic role to producers. The economic repertoire is occasionally mobilized to emphasize the necessity of adopting these measures on set, as we can see, for example, in the document noting that 'outdoor productions could often save 10–30% of fuel costs'. While economic costs are specifically highlighted as an incentive, on this level of operation, the assumption that sustainable means of production is a sensible approach seems to be almost taken for granted, with the rhetoric coming close to reaching a state of ecological habitus. Thus, there is no need to use the legal and the ethical repertoires, or even other frameworks to motivate or inspire different sections of the production crew as instead the focus is almost purely on core elements of production practice. We can see this in the ways the document frames the inclusion of materialities by encouraging productions to use waste vegetable oils that do not rely on destruction of habitats, or providing in-depth information on specific practical disadvantages and emissions benefits of hydrogen and methanol fuel cells. All of these contribute to a dynamic network where producer/managerial decisions are predicated on materiality. The consequence is an encapsulation of network dynamics much more in line with an ecocentric paradigm.

Table 5.7 Key words for *Factsheet 10: Sustainable Power for the Screen Arts on Location*

	Key words	Number of occurrences	(%)
Transactional	thinking	5	0.61
	arts	4	0.49
	events	4	0.49
	Media Greenhouse	4	0.49
	produce	4	0.49
	production	4	0.49
	rated	4	0.49
	screen	4	0.49
Total			4.04
Grounded	power	26	3.16
	fuel	23	2.79
	generator	12	1.45
	generators	10	1.22
	hydrogen	10	1.22
	biodiesel	9	1.09
	cell	9	1.09
	methanol	9	1.09
	sola	9	1.09
	diesel	7	0.85
	energy	6	0.73
	outdoor	4	0.49
	source	4	0.49
	box	3	0.35
	carbon	3	0.35
	electricity	3	0.35
	environmental	3	0.35
Total			18.16

(Media Greenhouse 2014).

These production documents allow us to go even deeper into the daily practice of the network. A memo for the crew of *Call the Midwife* (2014) provides an intriguing snapshot of how some of the top-down policies outlined in the management documents work on the ground level. Materiality is the key focus of this memo as it includes a list of actions from using recyclable bottles to car sharing, from water dispensers to uploading material on the internet. These are counterpointed with 'FACTS' about why these actions are necessary. The structure facilitates a productive dialogue between managerial and material considerations showing how both are interconnected to benefit from each other. As part of the mapping of the environmental media management network, this document is particularly useful as an indicator not only of translations of top-down directives into practice but also how material agency reverberates into organizational strategy and policy – the document provides an interface into the processes of the relationship between policy, management, practice and materiality. This is especially the case as the repertoires used to communicate these ideas combine both factual encouragement and inspirational rhetoric with a lighter tone than in many of the other policy and production documents.

To take a few examples, the memo uses puns and quips to communicate sustainability tactics. These are addressed directly at members of the crew who may share the codes on which this humorous take is based: 'Our sound recordist is going rechargeable – I know that is difficult to imagine given he never seems to move!' (Neal Street Productions 2014: 2). The document concludes with a statement combining the product brand with sustainability: 'Think of all the babies we will be delivering during this series – let's ensure they have a planet to grow up on!' (ibid.: 2). The informal tone of the memo uses typical strategies for internal corporate communications as it tries to mitigate any disruptive impact these strategies have on daily production practice. While the tone of the piece may appear somewhat informal, and the combination of material FACTS with the rhetorical style can be seen to simplify some of the material complexity of adopting sustainability on set, it affords a view into production processes in ways that refocus attention on the complexity of network dynamics at the interface of the grounded and the transactional, a balance also evident in the key word percentages (transactional at 9.09 per cent, grounded at 9.35 per cent) (Table 5.8).

If this document traces the reciprocal dynamics from the managerial levels to the material base and gestures to the ways they reverberate back to management, *Guide to Writing an Environmental Policy* by the consulting company Media Greenhouse makes this even more explicit (Media Greenhouse 2013). It provides an alternative angle to exploring network dynamics where production data based on grounded materialities is used to construct policy documents. Thus, the document highlights the ways management incorporates sustainability into their organizational strategy by responding to the imperatives set by grounded materialities. It is full of conventional management talk, utilizing approaches from corporate communication strategy to explain

Table 5.8 Key words for *Call the Midwife: Sustainability Memo*

	Key words	Number of occurrences	(%)
Transactional	fact	10	2.32
	production	8	1.86
	use	5	1.16
	Albert	4	0.93
	BAFTA	2	0.47
	controls	2	0.47
	department	2	0.47
	thing	2	0.47
	home	2	0.47
	TV	2	0.47
Total			9.09
Grounded	water	6	1.39
	green	4	0.93
	bottle	3	0.70
	car	3	0.70
	carbon	3	0.70
	couriers	3	0.70
	bins	2	0.47
	cars	2	0.47
	containers	2	0.47
	cup	2	0.47
	cycle	2	0.47
	eco	2	0.47
	food	2	0.47
	footprint	2	0.47
	sustainability	2	0.47
Total			9.35

(Neal Street Productions 2014).

how environmental sustainability policies should be set. For example, 'describe your starting point, set your goals, make the commitment' (ibid.: 1) here are the three categories for the management plan, which must be based on a pragmatic approach to policy: 'Set out measurable, achievable and realistic targets which are relevant to your company practices and time-scales' (ibid.: 1).

These vague, if well-worn encouragements, indicate some of the ways environmental sustainability is incorporated into management strategy – it combines the legal and the practical repertoires to certify the veracity of the plan in ways where the agentic concerns from grounded materialities are positioned in line with effective organizational management: 'a good environmental policy will be concise, easy to follow and implement, measurable and periodically reviewed' (ibid.: 1). Furthermore, the individual in charge of drafting it will need to answer 'who will take responsibility for achieving targets at each level of the organisation; how will you observe legal compliance' and have the document signed off by senior management (ibid.: 1). The approach taken here is much more in favour of a managerial approach as evidenced by the key words (transactional at 24.21 per cent, grounded at 4.84 per cent) (Table 5.9).

This fits in with management strategies used by companies to implement and measure their performance, and while there are no explicit grounded materialities featured in this document, they are part of its rhetorical scope as the material base is referenced explicitly in statements like: 'Define your largest environmental impacts (supported by data if possible). Use Albert to identify where these lie' (ibid.: 1).

Thus, both the *Call the Midwife* and the Media Greenhouse documents, despite their differences, can be considered practical attempts to manage the agency of grounded materialities and their influence on organizational strategy. These practical documents differ from the top-down documents in that there is no concrete emphasis here on costs or economic argumentation. The focus is much more on explaining practical matters in a way that makes sense to different parts of the production organization and which make the agentic presence of grounded materialities more tangible, felt, for the relevant parties. They show a more advanced stage of development from the rhetorical enticement and enforcement seen in many of the top-down policy arguments. In many ways, these documents demonstrate a progressive approach to environmental management that does not need to explain why sustainability is beneficial for the industry, but instead, they directly acknowledge that it needs to be adopted as a self-evident standard by all professionals in the industry. As part of consolidating the network, they testify to the reciprocal nature of dialogue between its nodes where both the eco- and the anthropocentric forms of rhetoric have a role to play.

Carbon literacy training at BAFTA

To date, we have evaluated a range of industry documents from the UK film and television sector. These documents have ranged from top-down

Table 5.9 Key words for *Guide to Writing an Environmental Policy*

	Key words	Number of occurrences	(%)
Transactional	policy	6	4.84
	make	3	2.42
	step	3	2.42
	targets	3	2.42
	stakeholders	2	1.61
	measurable	2	1.61
	provide	2	1.61
	set	2	1.61
	business	1	0.81
	cast	1	0.81
	client	1	0.81
	commitment	1	0.81
	communicate	1	0.81
	company	1	0.81
	compliances	1	0.81
Total			24.21
Grounded	Albert	6	4.84
Total			4.84

(Media Greenhouse 2013).

strategies to production manuals and reveal much of the dynamics at work in the industry. A frequent area of concern has involved balancing the eco- and the anthropocentric rhetoric in these management and production strategies in ways that environmental sustainability can merge productively with the core KPIs of the industry, especially as pertains to legal or economic incentives. However, these documents are not the only means through which the UK media industry constructs its network. Green Shoot holds annual industry training sessions for filmmakers and producers on integrating sustainability into their practice. Similarly, BAFTA hosts monthly sessions on carbon literacy training for the television industry. What follows is an observational study of one of these training sessions held in London in March 2017. Much of the session covers material that is available in most of the Albert literature. Yet, observational study and analysis of the organization and conduct of the session can reveal some of the power dynamics of the network, at least in terms of its paradigmatic eco- and anthropocentric orientation. The session provides key observational data on the actions of BAFTA in consolidating its networks, while it also abides with many of the key discursive practices described above, especially the consolidation of a particular reality that reflects the ways through which sustainability is adopted by these organizations.

The aim of the session, coordinated by television director and BAFTA board member Steve Smith, was to encourage the television industry to embrace environmental shooting practice as part of their everyday activities. In contrast to the emphasis on transactional managerialism in the industry documents, the session was clearly pitched at a more grounded level of rhetoric, and a more explicitly ecocentric approach. One of the first areas presented to the participants was a projected mood curve they were expected to undergo during the seven hours of the session. The shape of the curve starts out from an enthusiastic sense of participation in the event and slowly declines to extreme pessimism, only to rise again towards the end. The rationale for this expected fluctuation was that the first part of the session (2.5 hours) consists of a module on debates about climate change. This is a pathway fostered by the BBC Academy which designed the course as a means for staff members to adapt their practice to meet the challenges of climate change (BBC 2015: 18). The first half thus focused on fear rhetoric to shock the participants into acknowledging the realities of the threat the environment faces, including basic facts about the threat of climate change, such as the 2°C annual temperature rise threshold and the implications of the 450 ppm (parts per million) CO_2 concentration levels on the human body.

The rhetoric was constantly complemented by data from the IPCC with frequent emphasis on the notion that these are facts based on scientific evidence. Perhaps anticipating the easy dismissal of contrasting viewpoints as fake news, the presentation consistently indicated when something would be an opinion instead of a scientific fact. To emphasize the veracity of these facts, the presentation covered many key factors contributing to climate change, including the oil industry, farming and agriculture, manufacture, deforestation. To verify the seriousness of these considerations, arguments were complemented by information from the UK Climate Risk Assessment of 2017 and discussion of climate politics worldwide. The culmination of this section argued that the world has reached a tipping point with the runaway climate becoming an existential risk the industry must address. If the rhetoric for the carbon literacy session was quantified in similar ways to the policy documents, the volume of coverage of climate change would make this activity clearly ecocentric.

Accordingly, the focus on grounded materialities continued in the second half of the session which focused on sustainable practices in the media sector. The necessity to adopt environmental sustainability is communicated by simple figures to make the connections feel more tangible. One significant comparison presented was to do with the footprint of producing one hour of television content which equates to 14.6 tonnes of CO_2. This is, we were told, more than the 10.92 tCO_2 of the footprint of an average UK household. The visualization of the carbon footprint continued with a game called Play Your Carbs Right. Environmentally problematic acts like meat consumption or plane travel are converted into carbon equivalents which would be visually represented by how much CO_2 would fit into an average bathtub. This, again, provides an easy and memorable way for the participants to understand the impact of watching a film in HD streaming or a train journey to a film shoot on location. The ability to put a number on choosing air travel over train, for example, works as a productive instance of translating materialities into production management. Such visual means are designed to make the workshop participants reflect on their activities on set or in various phases of the production in ways that do not emphasize the transactional benefits of individual actions, but instead the grounded role they play. Thus, the workshop effectively translated management and production practices into grounded materialities, shifting the ways the industry management documents used transactional materialities – that is, coded translations of grounded materialities – to convince the readers of the benefits of environmental sustainability by using, for example, the economic and legal repertoires. Here, most of that work comes from the sheer weight of the grounded materialities, which create a reality where all actions on a shoot leave a material footprint that needs to be addressed.

This line of evaluation continued by exploring the lifecycle of objects and the circular economy, with the discussion focusing on the footprint of each facet of media production, including the sourcing of materials, manufacturing, transport, distribution and disposal. Practical tips were drawn from the *Graham Norton Show* that Steve Smith directs. The use of LEDs is one of the key strategies here as they can comprise up to 90 per cent of the potential energy reduction on set. They make the sets smaller as they do not need elaborate props and reduce the temperature on set but they also light the sets differently and require grading in post-production. The impact of LEDs on image quality has been a reason for hesitation in uptake as LEDs require changes in practice and shooting strategy, but as the discussion made clear, grading is now an industry standard and any difference in the image quality can be negated by standard post-production practices. The cost of LEDs is another problem that has hindered their adoption by the industry, but this is also changing as LEDs are now cheaper and becoming the norm in shooting practice. To apply this knowledge of the materiality of everyday objects to production practice, the group was tasked with identifying the origins of materials items such as a lipstick or a chair and where they end up after use. The roles of animal parts, minerals, chemicals, energy and labour of the human workforce are evaluated from a critical perspective to force us to rethink our daily and professional practices in a way that challenges business as usual and enforces us to reconsider our priorities. This sort of reflexive encouragement of the participants provides an exemplary way of integrating these individuals into the network.

Clearly, the aim of the training session was to encourage self-reflection on the part of the participants as the sessions, in general, go beyond the means available to the written documents in assimilating individuals into the network not only on the basis of their professional roles but also their private lives – which ideally translate into changes in professional practice. Furthermore, legal requirements and obligations were used throughout the session as another set of key incentives to urge the participants to join the network. The workshop discussions, accordingly, focused on the Paris Climate Accords, the UK's Climate Act of 1990 as well as the UK Climate Change Act of 2008, while other sections focused on staff recruitment and engagement. The potential of reducing legal risks was positioned as another incentive as these are due to increase as regulation across the different parts of the media industry solidifies. For one, the recycling of set and prop material is a legal obligation due to general waste laws and emissions from vehicles also need to meet legal requirements. The ways these were framed matter as while the transactional dimensions of regulation – the idea that sustainable practice is adopted because it will be a legal requirement – are part of this rhetoric, there is a sense here that legal imperatives are tangible, rational elements and not obstacles the industry has to meet. The necessity of incorporating sustainability thus emerges from grounded concerns and not the other way around, as tends to happen with some of the management documents.

These ideas were reinforced by a snapshot of the state of the television industry. All of the internal and externally commissioned activities at BBC and SKY have reporting as a mandatory prerequisite for funding. Thus, the workshop made it clear that if independent production companies want to work with the lead players in the industry, adoption of these practices is not a passing trend but an essential and soon to be mandatory part of the production infrastructure. These legal imperatives were enhanced by another group activity encouraging self-reflection on what exactly makes media a polluting force. The aim was to identify key sustainability items in statements by companies that have successfully branded themselves as environmental operators. These included SKY, Marks & Spencer and the NHS, all operations from very different backgrounds and with different core competencies. The aim was to provide a comparative discussion that reflects the ways different sectors have to tailor sustainability measures to meet their KPIs and abide by their particular regulatory environment. As instances of consolidating the connections of the network, these activities urged participants to explore their supply chain and consider ways to curb emissions through more sustainable procurement or commissioning practices. In contrast to the ways the documents emphasized managerial strategies to overcome the agentic role of grounded materialities, the focus here was much more on the fact that each managerial setting has a distinct material base that influences the operations of each company. Consequently, as with the legal requirements, the agentic role of the grounded materialities emerges as a principal driving force for the network being consolidated through these training activities.

The move from a restrictively top-down management to a more even level of participation was exemplified by another task based on considering the footprint of four different types of production: (1) location-based factual; (2) studio and location-based drama; (3) event broadcasts; and (4) studio-based production. Participants were asked to evaluate each type on the basis of office emissions, the implications for travel, on-set necessities, and post-production, with all the different production types categorized according to approximate CO_2 footprints (per hour of content) resulting from the specific conditions in which productions were done. These distinctions are significant as they indicate that there is no one-size-fits-all approach to conceptualizing a comprehensive strategy to cover the different areas and core competencies of each production type as, for example, studio or location shooting, as just one type of variable, will influence the footprint the production generates. After evaluation, the studio and location-based drama production came in at 54.3 tCO_2, location-based factual at an approximate 17 tCO_2, event programmes at 7 tCO_2, and studio shows at 3 tCO_2. These were translated into bath equivalents with drama the 'winner' at 108,500 baths of CO_2. The exercise was important to show not only the extent of the emissions from production activities but also the variety of approaches that are required to integrate environmental sustainability with television production practice. It served as a good reminder that media policy is complicated by the wide variety of practices under this generalized heading – there is no 'one-size-fits-all' approach. Legislating for all media under a single policy is not viable as an approach as any productive policy would need to rely on different ways of incorporating agentic grounded materialities into production planning – a notion we will address in later chapters.

The workshop acted as a practical balance to some of the official policy documentation which relied much more on the business KPIs of the media industry. There, the emphasis was often on management, not surprisingly, especially as many of the documents focus on relatively early stages of adoption, and thus have to use repertoires that make sense for industry players. The BAFTA training sessions are aimed at both management and crew and operate on a voluntary basis. Thus, the emphasis is on providing a set of transferable skills for all parts of the production infrastructure that enable the content of the workshop to flow into practice. It exemplifies the ways a network can be consolidated as internal strategy documents only work so far in scoping out means and methods of best practice. These need to be geared at those individuals who have enough power to institute sustainability into corporate practice or alternatively allocate funds to hire a green runner or an eco-supervisor. The carbon literacy training relies on dialogue from all parts of the production and facilitates a much more interactive environment. Thus, the role of different participants opens more space for agentic materialities to influence the constitution of the network.

CSR policies

The documents and practices we have explored to date focus on para-industrial considerations that position environmental sustainability as a viable industrial strategy for film and television. They reveal a complex network of relations between regulation, management, production and labour where organizations like BAFTA and Green Shoot facilitate a network that seeks to expand to new stakeholders. However, while these organizations operate in collaboration with other industry players across the supply chain, they form only a partial picture. Professionalized consulting companies specializing in sustainability management also play a major role in consolidating the UK environmental sustainability network, especially across different parts of the media industry. Our focus is on the consultation company Carnstone's work on environmental sustainability and its venture the Media CSR Forum (now updated to the Responsible Media Forum). They provide data and advice to the industry through workshops and consulting documents and collaborate with organizations like BAFTA, the BBC, *The Guardian* and Schibsted. Their operations are focused both on collecting data on emerging patterns in the industry and presenting this to network participants. The Forum has been operating since 2003 and has published a range of documents focusing on consolidating environmental sustainability as best practice. In combination, these documents show intriguing patterns of development in terms of understanding the impact of the industry. They are used here not so much as a set of documents talking about how the industry promotes itself para-industrially, but instead how these operations pitch ideas to external stakeholders and consolidate expansive networking practices. As they do so, they reveal much about the business side of adopting sustainability in the industry as well as of the role that ethical concerns play here.

Key CSR Issues for the Media Industry, the first of these documents, was compiled and published in 2005 in collaboration with KPMG (KPMG 2005). Produced as an advisory document for the industry, it is an early example of conceptualizing environmental prerogatives for the media. Much of the document is focused on corporate social responsibility with environmental considerations as a subsection of sustainability. This includes areas like promoting creativity, enabling freedom of expression, encouraging good citizenship and acting as a catalyst for community activity. These are all considered important positive steps for enhancing the responsiveness of an organization to societal problems. By surveying over 130 stakeholders from the media industry to the CSR community, from socially responsible investors to the financial community, the document provides a distinctively transactional approach to environmental sustainability. Crucially for this book, the data did not include 'internal stakeholders', i.e. employees or suppliers, but prioritizes consultation on the level of organizational management. Thus, this CSR document provides a contrast to the internal rhetoric of much of the other documents inspected here, especially to the practical focus of the BAFTA session, a notion very apparent from the frequency of key words (transactional at 18.51 per cent, grounded at 0.42 per cent) (Table 5.10).

In comparison to the industry documents observed above, the key word frequency is inherently oriented towards organizational management. Most of the terms revolve around organizational concepts and stakeholder priorities and meet the general KPIs of CSR. But in doing so, they reveal the minimal role environmental concerns hold even for stakeholder groups invested in corporate responsibility and sustainable development. The one area where the Forum does stake a specific claim to be specifically about environmental management is in its definition of the term: 'Maximising the positive and minimising the negative impacts of an organisation's operations and product output on society and the environment, by meeting stakeholders' expectations and complying with regulatory standards' (ibid.: 7). The description is a basic means of understanding the KPIs of environmental management of the media. It uses all four of the key repertoires – commercial, legal, ethical and practical – but does so in ways that positions these activities as a response to external requirements. The argumentation is about mitigating negative impacts that may arise from business or regulatory concerns. In doing so, this description is emulating the philosophy of sustainable development – as critiqued by Maxwell and Miller (2012) – in its framing of sustainability as an obstacle. Granted, this is an early example of the industry grabbling with sustainability in the media sector and reveals some of the limitations of its approach to the environment. As global warming and climate change have emerged as areas of concern in the public consciousness, the levels of knowledge about the power of the media to influence opinions on areas of environmental relevance have also increased. Thus, more calls for close attention to the content of these messages as well as their means of production have increased. It is not surprising then that organizations like KPMG and Carnstone have identified and started to provide specialist services in the field. This is, after all, a form of business and an expansive market.

Mapping the Landscape: CSR Issues for the Media Sector (Media CSR Forum 2008) continues the rhetoric of *Key CSR Issues for the Media Industry* as it aims to provide a more comprehensive charting of CSR initiatives in the sector. The survey on which the document is based was produced by Alcoa, an independent consultancy specializing in CSR activities, in 2008, and compiled by the Media CSR Forum to provide knowledge of key CSR issues driving the sector. Core issues concern the proliferation of digital operations and the ways they have changed the operational parameters of media companies. Driving this study is the suggestion that the updated data since 2005 show an increased interest and demand for the media to play a

Table 5.10 Key words for Key CSR Issues for the Media Industry

	Key words	Number of occurrences	(%)
Transactional	media	52	2.74
	CSR	36	1.90
	KPMG	28	1.48
	output	20	1.06
	industry	19	1.00
	information	15	0.79
	organisations	15	0.79
	stakeholders	14	0.74
	corporate	12	0.63
	organisation	12	0.63
	rights	12	0.63
	social	12	0.63
	community	9	0.47
	consultation	9	0.47
	cooperative	9	0.47
	international	9	0.47
	responsibility	9	0.47
	responsible	9	0.47
	society	9	0.47
	stakeholder	9	0.47
	standards	9	0.47
	content	8	0.42
	firm	8	0.42
	forum	8	0.42
Total			18.51
Grounded	environmental	8	0.42
Total			0.42

(KPMG 2005).

part in 'sustainable development and the mitigation of climate change' (ibid.: 4). To achieve these aims, its role is concerned with 'identifying areas for prioritisation, sharing best practice, engaging with stakeholders, identifying collaborative projects on key issues' (ibid.: 3). These areas are based on the industry's self-identification of its priorities for the 2005 study. For the 2008 document, a key difference is that climate change has now been separated from 'environmental responsibilities' (ibid.: 5) as it now appears as a general issue 'due to its recent prominence' (ibid.: 18). The text provides anonymous quotations from a workshop where the theme has been identified as a concern, yet the rhetoric only focuses on the communicative faculties of the media. The impact of the footprint does not register as a particular concern for media production at this point in time and appears as a CSR issue common to all sectors. As this lack of focus on emissions from the sector is based on the companies' identification of their priorities, it indicates the ways grounded materialities are often ignored as part of the realities of managerial and organizational strategy.

The marginal significance of environmental sustainability comes through in the core competency framework identified by the Media CSR Forum, as discussed in Chapter 1. Here, environmental considerations do not feature as part of the strategic or material foci of media organizations and are thus not identified as key priorities, nor are they seen as unique issues to do with the sector, but general considerations for most industries. The key word percentages support this view as they again highlight transactional patterns dominated by managerial considerations (transactional at 15.37 percent, grounded at 0.53 per cent) (Table 5.11).

In this document, grounded concerns are minimal compared to managerial rhetoric, and there is no real focus on particular emissions from media production, suggesting that at this level of networking, grounded considerations have little relevance. Only general approximations of sustainable development appear here, which makes sense considering that this is still a policy document aimed at outlining the appropriation of environmental sustainability into managerial strategy. As a consequence, CSR and its emphasis on sustainable development position transactionality as a key priority.

While these early examples are distinctly more focused on transactionality, the agentic role of different participants, including those of grounded materialities, in the network start to become more balanced by *The Media CSR Forum Activity Report* of 2012. This overview has more emphasis on explaining the particular environmental obligations of the media sector instead of focusing on providing information about general CSR concerns. According to the document, 'the practice of CSR and sustainability for media companies has many unique features that set it apart from other business sectors' (Media CSR Forum 2012: 1). While it provides brief explanations of different projects undertaken to understand the material impact of the sector, it does not go into much detail on what these material considerations are. The Forum has engaged with a range of media organizations including the BBC, BSkyB, Channel 4, the Guardian Media Group,

Table 5.11 Key words for *Mapping the Landscape: CSR Issues for the Media Sector*

	Key words	Number of occurrences	(%)
Transactional	media	160	3.78
	issues	70	1.65
	sector	41	0.97
	companies	40	0.94
	survey	35	0.83
	CSR	34	0.80
	important	27	0.64
	workshop	27	0.64
	respondents	26	0.61
	forum	25	0.59
	issue	24	0.57
	information	23	0.54
	validation	21	0.50
	change	20	0.47
	development	18	0.42
	stakeholder	18	0.42
	corporate	17	0.40
	social	17	0.40
	advertising	15	0.35
	diversity	15	0.35
	integrity	14	0.33
	content	13	0.31
	output	13	0.31
	community	12	0.28
	investment	12	0.28
Total			15.37
Grounded	climate	15	0.35
	environmental	12	0.28
Total			0.63

(Media CSR Forum 2008).

Alma Media, ITV, News International, Pearson, Random House, and Reed Elsevier and indicates awareness of the different types of impacts these organizations have.

The most intriguing part of the document is the inclusion of a substantial focus on the brainprint, a concept not fully addressed in any of the earlier documentation. This shifts focus away from the material impact of the industry and provides further support to the argument that the media's environmental costs are much lower than those of heavy industries, and their core competencies are about communicating knowledge. Simultaneously, as the document argues, environmental management is a necessary action for any responsible media organization. Yet, the ways this is addressed in the ordering discourses of the CSR network remains an issue, as evident in the clear transactional focus of the key words (transactional at 23.23 per cent, grounded at 1.10 per cent) (Table 5.12).

Of course, it would be unfair to require short summaries like the *Activity Report* to contain in-depth information on all aspects of the organization's operations. But simultaneously, the emphasis on management indicates the persistent framing of environmental materialities as a business concern and a management challenge. In contrast to the levelling of power between grounded and transactional materialities emphasized in many parts of the film and television sector, these CSR documents seem to emphasize that the Media CSR Forum views sustainability as a potential challenge and a business opportunity.

Despite these percentages, it would be overstating the case to suggest that the Media CSR Forum consciously downplays environmental areas in favour of more managerial concerns. Rather, this is a result of the company's role as a consultancy using industry data to construct an overview of the industry's approach to environmental management. The aims of the Media CSR Forum are to identify industry KPIs and provide strategies that balance sustainability with operational priorities. Thus, it is not entirely surprising that their approach fits more into an anthropocentric paradigm. This balancing is further exemplified by the 2013 document *Does It Matter: Material, Strategic or Operational?* – a study by the Media CSR Forum to evaluate whether sustainability is in fact a prime concern for the sector at all. It starts with a very revealing statement from James Weatherby, the Chair of the Church of England Ethical Advisory Group:

> The Media CSR Forum is to be congratulated on this clear and concise report on the social sustainability issues facing the media industry. The report is penetrating in its analysis of the extent to which these issues matter to the media industry, and therefore to its investors, from a financial perspective. The disappointing but realistic conclusion is that – under current models at least – they do not matter very much.
>
> (Media CSR Forum 2013: 2)

Table 5.12 Key words for *The Media CSR Forum Activity Report*

	Key words	Number of occurrences	(%)
Transactional	media	39	4.74
	forum	19	2.31
	report	16	1.94
	issues	14	1.70
	CSR	9	1.09
	members	9	1.09
	Carnstone	8	0.97
	companies	8	0.97
	stakeholders	8	0.97
	LLP	7	0.85
	partners	7	0.85
	secretariat	7	0.85
	group	6	0.73
	meetings	6	0.73
	sector	5	0.61
	activities	4	0.49
	brainprint	4	0.49
	company	4	0.49
	debate	4	0.49
	impacts	4	0.49
	published	4	0.49
	research	4	0.49
	strategy	4	0.49
	business	2	0.25
Total			23.23
Grounded	environmental	5	0.61
	sustainability	4	0.49
Total			1.10

(Media CSR Forum 2012).

The document lays out the tripartite distinction explained earlier in this book concerning strategic, operational and material priorities for the industry. As explained:

> A material issue is financially significant over the short to medium term, i.e. it has the potential to affect a key financial indicator, e.g. profits or revenue, by around five per cent or more within a two-year time period.
>
> A strategic issue has the potential to significantly affect the ability of the company to deliver its strategy in the medium to long term.
>
> An operational issue matters for other reasons – internal, reputational, efficiency – but is neither material nor strategic. Under normal circumstances, it does not represent a significant threat to the company.
>
> (ibid.: 5)

This framework suggests that the footprint of the industry is not a significant concern due to the ways it links with the media sector's KPIs of producing content and delivering it to audiences. As argued, environmental management is only an operational concern and has relatively little significance for the bottom line. The Media CSR Forum thus suggests that this insignificance is the reason why financial and regulatory concerns are unlikely to lead to real change in the industry's environmental performance. For them, any change is likely to be a consequence of individual motivation. Again, the key word frequencies emphasize the disparity between the transactional and the grounded rhetoric, suggesting that consulting operations tend to frame environmental concerns as a management problem (transactional at 22.89 per cent, grounded at 0.80 per cent) (Table 5.13).

Continuing from this rather pessimistic view, in 2015, the company published an *Annual Review* that makes the case for business a lot more explicit than it has been in the earlier work. According to it, the Forum aims 'to define and develop responsible business practices for the media sector' (Media CSR Forum 2015: 2). The aim is to provide a service that communicates environmental sustainability issues to a wide range of participants but does so in a framework emphasizing business considerations, especially as the Forum works with 'investors probing into specific areas of media sustainability' (ibid.: 3).

The balance between the managerialist and the more grounded sustainability incentives is more refined here than in earlier publications. The document provides far-ranging discussion of different types of footprints with particularly advanced steps for 'understanding the environmental impacts of digital content' (ibid.: 4). This is a significant section as it shows the ways industry initiatives can be at the forefront of developing knowledge on emerging areas with substantial environmental impacts, including the role of digital communications and technology. For example, it incorporates data from Greenpeace and other environmental NGOs to suggest that 'in a broader context, data centres account for around 2 per cent of global emissions' (ibid.: 4). To address these considerations, '2015 saw the launch of a

Table 5.13 Key words for *Does It Matter: Strategic, Material or Operational?*

	Key words	Number of occurrences	(%)
Management	media	134	4.09
	companies	75	2.29
	sector	42	1.28
	operational	40	1.22
	issues	39	1.19
	material	38	1.16
	strategic	37	1.13
	forum	29	0.88
	CSR	26	0.79
	content	24	0.73
	editorial	22	0.67
	company	19	0.58
	access	18	0.55
	output	17	0.52
	report	17	0.52
	market	14	0.43
	affect	14	0.43
	diversity	14	0.43
	sectors	14	0.43
	social	14	0.43
	staff	14	0.43
	investment	13	0.40
	ownership	13	0.40
	public	13	0.40
	transparent	13	0.40
	impact	12	0.37
	markets	12	0.37
	society	12	0.37
Total			22.89
Material	environmental	13	0.40
	responsible	13	0.40
Total			0.80

(Media CSR Forum 2013).

Forum working group to explore the carbon impacts of digital media content' (ibid.: 5). Yet, this innovative view is counteracted by the inclusion of arguments from Craig Bennett from Friends of the Earth suggesting that the brainprint remains the most significant aspect of the relationship between media and sustainability. Thus, while the document incorporates a range of materialities into these ordering discourses, their positioning is arguably compromised by both an emphasis on transactionality and the use of alternative frames such as the emphatic focus on brainprint, as reflected in the key word frequency (transactional at 17.32 per cent, grounded at 1.91 per cent) (Table 5.14).

Conclusion

Concluding this section with a focus on CSR is necessary to highlight the extent to which the industry considers organizational management to be the essential strategy for incorporating environmental sustainability protocols into its operational parameters. And indeed, we should not underestimate the considerable contribution of such strategies to the overall 'greening' of the industry. After all, these initiatives are in the early stages of development and implementation, and consolidating a functioning network is a key priority to get the industry to take sustainability seriously as a legal, ethical, practical and commercial imperative. Simultaneously, the ways these strategies are communicated to potential participants brings up questions over framing and the targeting of audience demographics. There is a clear emphasis here on management and business concerns, which results in the consolidation of these parts of the network through less emphasis on grounded materialities than was the case with many of the film and television organizations where the focus is often on practical matters, and thus on a more agentic role for grounded materialities. Simultaneously, the role of the Forum is about consolidating an organizational management network and, thus, using the appropriate repertoires is necessary – despite how an ecocritical perspective may consider them as lacking in substance.

One of the key challenges of any such network involves overcoming the industry's lack of awareness of its footprint and the costs associated with sustainability. Due to these challenges, the rhetoric used by sustainability initiatives must deploy a combination of transactional and grounded imperatives for adopting green practice. This makes sense as the media are a capital-intensive and often explicitly commercial venture, necessitating that these ordering discourses speak the language of commerce and management as well as ethics and practice. Through this, the discussion has emphasized the ways the eco- and the anthropocentric paradigms struggle for balance in the constitution of the relatively successful case of the UK environmental management network. These contradictions and challenges indicate some of the key areas of focus for a critical perspective on environmental management of the media, including the role of legislation established at different institutional levels based on diverse means of governance, to which we now turn.

Table 5.14 Key words for *Annual Review*

	Key words	Number of occurrences	(%)
Transactional	media	45	3.30
	forum	33	2.42
	companies	18	1.32
	report	16	1.17
	content	13	0.95
	annual	12	0.88
	social	11	0.81
	impacts	8	0.59
	meetings	8	0.59
	stakeholders	8	0.59
	mobility	7	0.51
	participants	7	0.51
	Carnstone	6	0.44
	CSR	6	0.44
	digital	6	0.44
	strategy	6	0.44
	world	6	0.44
	conference	5	0.37
	experts	5	0.37
	group	5	0.37
	hosted	5	0.37
Total			17.32
Grounded	sustainability	10	0.73
	change	6	0.44
	climate	5	0.37
	environmental	5	0.37
Total			1.91

(Media CSR Forum 2015).

Bibliography

BAFTA. 2015. *Environmental Sustainability: Ready Telly Go!*London: BAFTA.

BBC. 2011. *Albert – Programmes that Don't Cost the Earth*. London: BBC.

BBC. 2015. *The Environment: A Sustainable Approach*. London: BBC.

BFI. 2015. *Trialling the BS 8909: Specification for a Sustainability Management System for Film*. London: BFI.

BSI. 2014. *BSI Case Study: The British Film Institute*. London: BSI.

Greening the Screen. 2015. *Inspiration and Information for a Sustainable Screen Art Industry*. London: Greening the Screen.

KPMG. 2005. *Key CSR Issues for the Media Industry*. London: KPMG.

Maxwell, Richard and Miller, Toby. 2012. *Greening the Media*. Oxford: Oxford University Press. Media CSR Forum. 2008. *Mapping the Landscape: CSR Issues for the Media Sector*. London: Media CSR Forum Secretariat.

Media CSR Forum. 2012. *The Media CSR Forum Activity Report*. London: Media CSR Forum Secretariat.

Media CSR Forum. 2013. *Does It Matter: Material, Strategic or Operational?* London: Media CSR Forum Secretariat.

Media CSR Forum. 2015. *Mirrors or Movers*. London: Media CSR Forum Secretariat.

Media Greenhouse. 2013. *Guide to Writing an Environmental Policy*, London: Media Greenhouse.

Media Greenhouse. 2014. *Factsheet 10: Sustainable Power for the Screen Arts on Location*. London: Media Greenhouse.

Neal Street Productions. 2014. *Call the Midwife: Sustainability Memo*. Chertham: CTM Productions.

6 Regulatory infrastructure

Introduction

Environmental management needs to be perceived as a power network if we are to fully understand the complex means through which its different constituents influence one another. The ways these networks are constructed and maintained via ordering discourses highlight the balance between an eco- and an anthropocentric worldview and indicate some of the vested interests at play in instituting sustainability as an area of concern for the industry. The UK network discussed in Chapter 5 is a successful manifestation as it consists of multiple participants and expands internally and externally in various directions. Grounded and transactional materialities hold various forms of agency as part of this network but ultimately act as key ingredients tying the nodes of the network together. Simultaneously, we must be mindful of the complexity of these networks, adopting what Law (2007) suggests as an essential facet of actor-network theory, which is its focus on the multiplicities and fluidities of how networks form and transform. To comprehend these overlaps and intertwined power relations, it is especially important to place these networks in their wider ecology. This complexity suggests that the network is not some stable or unified conception, but in fact comprises complex entanglements, embedded in various power relations. These include the role of international and domestic regulatory regimes, ministries, non-governmental organizations (NGOs), media consultancies and production organizations. Once we add the material environments of specific contexts into the mix, all with different geographical and environmental bases for resources, as well as differing approaches to addressing them, we arrive at an understanding of the complexity of agency involved in all these networks.

The central role of various forms of energy – be they human or natural resources – in networks suggests energy holds a considerable agentic role in the relations between regulation, production, practice and management. This is especially significant when we consider the role of policy and regulation in these networks. As we have argued above, the use of managerial discourses tends to tip the balance of the network in an anthropocentric direction. This reflects the pragmatic realities of business operations but can also lead to active marginalization of environmental concerns. To attain a more ecocritical perspective, we need to approach media policy not only as a managerial tool but also a means for grounded materialities to influence the constitution of the network. Thus, policy can act as a buffer zone against more anthropocentrist interests that consider sustainability either as an obstacle or as an obligation they have to meet. Accordingly, we must turn our attention to the regulatory environment as regulation is in a unique position to both limit the agentic potential of grounded materialities as well as enable them to find more potential to influence network dynamics. Finally, policy is key to establishing the reality of the network – that is, its conceptions of appropriate behaviour and best practice, as well as its attitudes and values in relation to environmental sustainability. Furthermore, policy can allow us to focus on the cultural and political environment in which regulations are set as they respond to the particular requirements that are imposed on media production in any specific context due to a variety of cultural, political, environmental and social factors. It is to this context we now turn.

The Nordic countries

To conduct this analysis, the second half of the book will focus on a set geopolitical and cultural context – that of the Nordic countries. Chapter 5 has outlined the construction of a successful media management network in the UK but attempts at similar networks have not conjoined in such a productive way in the rest of Europe. These aspiring networks need to be studied to understand the obstacles and complications that impede the formation of productive networking, and as a consequence, of a feasible approach to developing environmentally sustainable media management. At the time of writing, there simply does not exist any real governmental policy on environmental sustainability and the media, requiring that if these initiatives are to be adopted they must be achieved through industry self-regulation. This approach has functioned well in parts of the UK network but less so in other contexts. Yet, to understand the lack of clear self-regulation in, for example, the Nordic countries, we must provide a sufficiently wide-ranging analysis of the regulatory environment, including analysis of intergovernmental and supranational regulatory systems as well as regional and national policies. To do this for all 27 member states of the EU, for example, would be beyond the scope of this study. Thus, our focus will be predominantly on conducting a geoculturally located study on the role of sustainable media management in the Nordic countries.

The Nordic countries, both as a set of small nations and as a regional constellation, are perceived as some of the greenest in the world. They often pride themselves on their green credentials and global responsibility. From summits on climate change to debating resource and energy production, from sustainable development policies to enhancing the greening of urban spaces, the Nordic countries maintain leading roles in environmental progress and policy on an increasingly global level. Christine Ingebritsen explains the principles of these areas well:

> As in other areas of foreign policymaking, each Scandinavian government has developed a niche in climate policymaking, and they collectively seek to shape the direction and substance of global collaboration. Scandinavia is a site of entrepreneurship, where new technologies and ideas are put into practice and exported to the world community thereby contributing to the greening of capitalism. Scandinavia, in contrast to other larger societies, has implemented climate adaptation at a faster rate and greater intensity for reasons associated with historic dependence on finite resources and national institutions that govern political economy.
>
> (2012: 88)

This optimism is echoed by Solability's Global Sustainability Index where all five of the Nordic countries led the poll for the fifth year running in 2015. Similarly, the same organization's Social Cohesion ranking is headed by the same countries. Their GDP per person is higher than the EU average (Norway, Sweden and Denmark are the top three) and they all, except Norway, spend more on R&D than the EU average of 2 per cent of the GDP. Norway spends 1.7 per cent but the total tends be higher due to Norway's high GDP. Due to the availability of natural resources, hydro-electric and geothermal power are major sources of energy in the Nordic countries. Since 1990, there has been a rise in the proportion of renewable energy in the overall energy supply of the Nordic region. The most substantial increase has occurred in Denmark while Norway has seen a decrease from nearly 53 per cent to 42 per cent. While advances are clearly being made, other areas are still lacking. For example, environmental taxes show a downward trend in Denmark and Iceland as well as a modest rise in Finland and Sweden. Finland and Iceland deposit a lot of waste into the environment whereas the other countries have made more efforts to enhance energy recycling. While GDP has grown, so have overall energy consumption and greenhouse gas (GHG) emissions (Nordic Co-operation 2013).

Clearly, there is a case to be made for the leading role the Nordic countries maintain in both social welfare and sustainability. Yet, it is the approach to environmental sustainability that emerges as a problem here, based as it often is on the ideologically conflicted concept of sustainable development. These societies are characterized by a paradoxical combination of neoliberal practices and welfare principles. In fact, the notion of the paradox is a useful metaphor to employ here as social and environmental sustainability in the Nordic countries is fraught with tensions and contradictions. These societies often pride themselves on being international leaders and role models in many areas of environmental activity, often for good reason. Simultaneously, much of the economic and resource infrastructure of countries like Norway and Finland relies on resources such as oil and nuclear power.

Energy politics, to take just a few examples, are rife with ideological and practical contradictions from the pervasive corruption scandals of Norway's Statsoil to the recent corruption allegations and construction failures in Finland's nuclear energy production infrastructure. At the same time, the global proliferation of the Nordic model is often reliant on a triumphant declaration of the exceptionality (Browning 2007) of socio-industrial organization of these welfare societies as well as their simultaneous acceptance and support of neoliberal tendencies. Accordingly, designations such as 'welfare capitalism' characterize the framework that shapes the media cultures of the region and influences any environmental strategies they adopt.

These contradictions are evident not only in environmental policy, but also in the organization and practice of the media industries. They enjoy substantial legislative freedom as well as financial and organizational support, yet environmental regulations are not thoroughly implemented into their practical management, even if environmental concerns are frequently given voice in the media. These assertions are supported by the lack of reference to environmental issues in media research. Not only is the carbon footprint of the Nordic media industries practically ignored in academic research, but the very real stakes of the local and global impact of the industries (both in terms of messages conveyed and the resources used) are not comprehensively understood by regulators or the industry. We need to focus on contexts such as the Nordic countries, as while it is clear that environmental issues flow over the borders of nations, often deliberately, especially in cases such as the recycling of e-waste generated by the information economies of the 'developed world', the structures and directions of this flow necessitate that critical focus lands on the affluent contexts that benefit from this flow, and who often use the said media industries to project entirely different impressions of their environmental role.

This chapter will address the role of regulation in facilitating (or in many cases, impeding) the construction of a sustainable media network. As we saw in the case of the UK, governmental legislation and infrastructure play a part as facilitators of the network. These are often attained through large-scale climate goals such as the Climate Change Acts in England and Scotland. Yet, the real onus relies on the sector with publicly funded organizations like the Arts Council England and BAFTA mandating environmental responsibility across their collaborative networks. One of the reasons to focus on the Nordic countries is that these sorts of incentives by central cultural institutions do not exist in this context as of yet, and it is thus important to address the political and cultural reasons for these omissions.

There are two key arguments explaining this lack of regulatory interest. The first is to do with the complexity of the regulatory infrastructure, or what I call the responsibility deficit. The Danish Ministry for the Environment, for example, suggests that any prerogatives for media legislation would need to come from the EU (pers. interview, 28 April 2015). Without delving too far into the rationale for this argument which will be addressed below, the perspective indicates a lack of understanding of the diversity of roles media play in society, including the necessity of curtailing some of their more negative effects. The other argument comes from Denmark as well and is premised on the argument that these societies are already executing wide-scale reductions in harmful emissions in all aspects of society and media will inevitably be swept along. Thus, responsibility is, in reality, left to the industry, but as there are no real functional obligatory passage points (OPPs) in the Nordic countries such as the Arts Council England and BAFTA performing crucial roles in establishing and sustaining the network, their absence fundamentally undermines the establishment of sustainability as an industry norm. One of the key aims of this chapter is to initiate discussion on these omissions in the infrastructural regulatory network and chart the impact these disconnections have for the industry.

This chapter outlines the existing provisions for the environmental regulation of the media from, initially, the perspective of legislators and institutes. The data are premised on both policy documentation and a set of comprehensive interviews with a range of Nordic civil servants and professionals. The sources for the data consist of governmental ministries, regulatory authorities, environmental organizations, business authorities and media companies. We initially take a step back from the ways the industry conceives of its own environmental role and focus on some of the regulatory frameworks that facilitate and necessitate the work we discussed above. Providing this level of contextual information is important to comprehensively analyse where and how the sustainable media networks emerge. For example, the UK laws on green energy and sustainability are lacking when compared to the Nordic countries, evident in the respective carbon divestment goals and regulations and incentives on areas like vehicle emissions and commitment to renewable energy resources.

An example of this can be seen in the Danish Government requiring reporting of corporate responsibility policies under the Danish Financial Statements Act. These apply to large companies with a balance sum above 19 million EUR, revenues above 38 million EUR and more than 250 employees (Kauffmann et al. 2012: 18). In the UK, the Environment, Social and Governance rules under the Companies Act 2006 specifies that reporting is necessary for all listed companies. Both of these regulatory frameworks exclude many media companies from being obliged to report their emissions, especially as the industry tends to revolve around small to medium-sized enterprises. There is an important distinction here between the UK and the Nordic contexts as in Denmark, even public broadcasters like Danmarks Radio have to provide reports on their operations, whereas the UK laws would not apply to non-listed companies like the BBC (though other laws apply to areas such as office operations and recycling). Larger media companies like Sky or the publisher Schibsted would need to report on their carbon emissions under these regimes. According to the carbon training session by BAFTA, the media industry in the UK compensates for the lack of infrastructural regulation by policing its own responsibilities. Nordic regulators see it from the opposite perspective – the lack of media regulation emerges directly from the strong level of infrastructural legislation in other environmental areas. The focus on regulation is necessary, then, to understand how environmental management often operates as an outcome of a specific cultural and political environment.

As we have established, the Nordic countries have relatively comprehensive environmental regulation in place yet this does not apply to the media in any binding way. The initiatives that exist are minimal or formulated by public broadcasting institutions like the Finnish National Broadcasting Company or Sveriges Television. The roots of these omissions can be traced to legislative oversight. Sauli Rouhinen from the Finnish Environmental Agency (pers. interview, 20 October 2014) suggests that all the environmental regulation that Nordic countries implement are developed from the basis of the EU. The implication is that were media-specific legislation to be developed, they would have to be led by EU incentives. These developments would have to go through the consulting and legislative process for the establishment of environmental laws and initiatives across the EU as a whole and be designed by focusing on media-specific key performance indicators (KPIs). The problem is that while the EU has enforced implementation of ICT directives such as the Waste Electrical and Electronic Equipment Directive (WEEE) and Regulations on Hazardous Substances (RoHS), they do not have a concrete policy on regulating other parts of the media industries, especially the production and distribution of media that have a much wider footprint than is covered by WEEE or RoHS, such as film production or digital publishing.

One of the obstacles facing environmental media regulation is, as Christian Toennesen (pers. comm., 25 May 2015) suggests, that the same laws apply to them as pharmaceuticals and other dissimilar sectors, that is, they are regulated on the basis of the scope of their production operations. This is based on an established EU legislation precedent for legislative oversight that necessitates large companies (those with more than 50,000 employees) to report on their footprint. The obligations apply to many key media organizations, from broadcasters like the BBC to publishing companies with an international range of operations who have to meet international and domestic regulatory demands. While they have to report on their footprint, they are not regulated on the scale or type of emissions emerging from the specifics of their core business. Rather, they are considered like any unspecified large company operating – as well as owned – internationally. This leads to the marginalization of environmental sustainability as an operational concern, and as such, it is not of substantial value for corporate strategy. Yet, the film and television sector is acutely aware of potential regulatory oversight as the Greening Film website (2017) suggests: 'European legislation will have a growing impact on the way businesses plan for environmental sustainability.' To counteract such developments, the European Regional Development

Fund allocated 2 million euro to a venture called Green Screen in 2015. This hosts nine partners from different parts of Europe and aims to reduce the carbon emissions of the film and television industries. As these strategies develop, more incentives will appear for sustainable media productions. But, arguably, it is only widespread legislative 'encouragement' – such as that employed by BAFTA and the BBC – that results in sustainability emerging as a standardized approach.

Furthermore, if the significance of EU governance is a prerequisite for any development of media legislation in the Nordic context, the industry's focus on the brainprint poses additional complications for addressing its footprint. For example, the European Economic Area Report, *Communication, Environment, Behaviour* (European Environmental Agency 2015) acknowledges the necessity of integrating communication strategies as part of environmental policy. Generating awareness is a means to improve implementation and ultimately contribute to facilitating a transition to a resource-efficient economy. Such strategies are vital for generating overall impact on the environment as the core competencies of media are here harnessed to support policy development. Used alongside other policy tools, communication can prove a very efficient – and cost-effective – policy tool, as *Communication, Environment, Behaviour* argues. These approaches fit with general perceptions of the industry's environmental role, but, at least for our purposes, they are lacking in scope as they do not consider the material role of the media. Instead, the industry is only seen to have instrumental value.

Environmental regulation in the EU

To understand the rationale for the lack of focus on the industry's material role and to contextualize sustainable media management appropriately, we now turn to exploring the history of environmental regulation in the EU. This history is significant for understanding the context from where sustainability initiatives for the media emerge. They reflect the general reality that facilitates the emergence and existence of the network as well as setting the parameters for the use of eco- and anthropocentric rhetoric. Unpacking this history is vital to identifying how other key stakeholders view media regulation in relation to general environmental frameworks, including its key aims and responsibilities. They also contribute closely to both the more advanced environmental policies in the UK as well as their absence in the Nordic countries.

Ever since the conference on the Limits of Growth in 1972 focused political attention on environmental issues, environmental regulation in Europe has undergone seven cycles of Environmental Action Plans (EAPs) (Hey 2005). These EAPs can be considered as formulating the organizing discourses for the infrastructural network in which the media network inevitably operates. Each action plan comes with a particular set of regulatory developments that respond to contemporary concerns in the environmental sciences and public opinion, all of which translate material considerations into industry protocol and management practice. These protocols would often follow contemporary requirements for environmental policies. According to Hey, 'procedural requirements, framework directives, voluntary agreements and self-regulatory information and management tools' (ibid.: 24) were key parts the first EAP cycle had to establish. These generalized protocols relied on instruments that are 'consensus-oriented and require the co-operation of industries'. The emphasis on co-operation and consensus is already here established as a norm for environmental regulation, a factor intriguingly repeated in many aspects of sustainable media policy. As with the EAP, dialogue across a wide range of stakeholders in the media industry was consolidated as the necessary initial step for industry organizations like the EMA and the UK Film Council, who provided early guidelines for adopting environmental sustainability in the sector. While the strategies of these organizations do not comprise comprehensive policy in a similar sense as the EAP efforts, they can provide incentives that are eventually consolidated as normalized protocols adopted by the industry. Thus, interestingly, from early on, the development of media policy reflects many of the historical developments of environmental regulation, a pattern that provides a comparative basis of analysis for understanding the history of sustainable media policy in the Nordic countries.

Another key area where media policy reflects the development of EAP plans is in the importance that national regulatory frameworks hold. The balancing of national and supranational forms of governance underlies the debates over the Third EAP (1982–1986) and the Fourth EAP (1987–1992). They gesture to many reasons for the pervasive reliance on intergovernmental regulations that are frequently mentioned as justification for overlooking the need to establish media-specific policy through domestic cultural or environmental institutions. A key area of these EAPs concerns the lack of unilateral agreement on emissions standards between domestic regulations and the EU market. This has been a traditional cause of friction for internal trade inside the bloc and, thus, a process of collaborative harmonization was required as 'environmental emissions standards needed to be harmonised to avoid distortions to industry competitiveness. Product regulations had to be harmonised to avoid non-tariff barriers emanating from different national product norms' (ibid.: 20). Thus, emphasis was laid on equalling differences and projecting a common framework for key environmental concerns. While each country has considerable room in setting their own benchmarks under this framework, many of the key regulatory decisions would be coordinated intergovernmentally and, by extension, the development of policy in new areas such as the media would require incentives as well as policy dictation from the EU. Such moves have laid the foundation for the lack of an established environmental regulation framework for the media as the footprint of the sector would, first of all, need to be of sufficient scope to emerge on the radar of EU regulation, and subsequently, be adopted to the specific environmental and regulatory conditions of each nation.

The Fourth EAP also highlights an aspect that has become a dominant facet of sustainable media policy – the role of financial considerations in establishing and consolidating regulation. This approach can be characterized as ensuring that environmental considerations are perceived as 'an integrated part of economic decision-making' (ibid.: 20). The approach here has been on analysing the impact of key economic sectors on the environment. The media have not played a part at this stage, of course, but the emphasis on sectorality is an opportunity for more specific regulation, precisely of the type that would be able to meet the core competencies of the media industry. This approach takes into account the key operating principles of each industry and aims to target them instead of emphasizing often abstract quantifiable limits that companies must exceed to be considered a climate threat. A problem with this approach, specifically relevant to the media, is that it often excludes those industries with a comparatively smaller footprint than the heavy industries. Furthermore, the emphasis on enticing cooperation, prioritizing pragmatism, identifying key stakeholder priorities and a general foregrounding of economic logic reflects the consolidation of the concept of sustainable development by the Brundtland Commission (World Commission on Environment and Development 1987). To quote the well-known outline, this involves a form of 'development that meets the needs of the present without compromising the ability of future generations to meet their own needs', a definition which has frequently been critiqued for its anthropo-centric logic and one which, again, resonates with key problems within the infrastructure of the sustainable media management network.

The concept of sustainable development has gradually become a normative reference for environmental policy in the EU. For example, it was included as one of the key principles of the Treaty on European Union of 1997. The objective of this Treaty was to identify areas where environmental and economic areas would benefit from mutual integration. The shift to framing environmental activity as 'sustainable development was perceived as a tool for improving the state of the environment, social efficiency and competitiveness simultaneously' (Hey 2005: 21). Consequently, regulatory frameworks were modified in favour of a shift towards economic and fiscal instruments designed to transform stakeholder behaviour. The result of this paradigm shift was a loosening of regulatory means in favour of mechanisms that played up the economic over the environmental and left the responsibility of adoption to the organizations: 'the new regulatory approach fitted better into the "neo-liberal wave" rather than the previous command-and-control approaches, because it focused on market mechanisms, deregulation, and self-regulation' (ibid.: 21). Sustainability was thus distanced from the more obvious control mechanisms of the earlier periods, reflecting the increased emphasis laid on market-based tools.

In many ways, this reflects the development of the sustainable media network transitioning from voluntary mechanisms to addressing schisms between international and domestic regulations. While financial considerations play a part in even publicly funded operations such as the BBC, they are much more central in the establishment of CSR operations. For example, the work of the Responsible Media Forum continues to be premised on very similar principles highlighting explicitly commercial benefits. Another consideration is that the EU countries are regulated under OECD rules, whereby economic argumentation finds an even more solid base due to the general priorities of the organization. An OECD document on Innovation and Transfer echoes some of the developments of framing sustainable media policy as an economic incentive in suggesting that 'economists have long argued that directly changing the relative price of polluting inputs (or products) through market-based instruments, such as taxes or tradeable permit schemes, is the most effective way to induce innovation' (OECD 2011: 14). Simultaneously, many consider such carbon trading schemes a panacea that substitutes financial instruments over real gains in emissions reduction. These mechanisms are not part of the arsenal of media organizations, though whether this is due to the relatively limited scale of emissions of the media industry or the financial requirements of these measures is open to debate. While the delineation between market-based instruments and more direct forms of regulation is heavily contested, the use of these tools makes a productive point about sectoral differences. For sectors like heavy industry or energy production, carbon trade or other economic schemes may work well. For smaller-scale emitters, like cultural industries or the media, these practices would not make much financial sense. Here, more voluntary incentives or regulation, both with an emphasized financial bent, may be more productive.

The strategic manoeuvres of policy reflect once again the logic adopted by the media industry, showing us how sustainability policies face many similar development obstacles regardless of the context or the sector of the industry. Once environmental strategies are positioned as part of sustainable development, they are characterized by similar advances like the Fifth EAP's emphasis on more market-oriented instruments including both voluntary and fiscal incentives. Similarly, we see that organizations like the BBC require extensive reporting while others such as the EMA emphasize PR and reductions in costs. This indicates how a strategic shift from imposed management to voluntary agreements and the use of self-regulatory information and management tools has become the dominant pattern underlying policy development. Yet, even as these developments proliferate, voluntary mechanisms have to be sold to their potential 'users'. The responsibility for this often lies with particular national cultural institutions who facilitate the means for organizing the participation of companies as part of their network.

Yet, this emphasis on the national institutions brings us back to the uncertainty over allocating responsibility for developing policy. Jordan and Adelle suggest in their edited collection *Environmental Policy in the European Union: Contexts, Actors and Policy Dynamics* (2012) that the implementation of regulations continues to be the most contentious aspect of EU governance. Saskia van Holten and Marleena van Rijswick (2014) argue similarly that while the EU has a set of Framework Directives, administering and implementing them form the responsibility of individual Member States. This is a key question for our work as it focuses attention on the ways each of the Nordic countries sets targets for their emissions. The strategies of each country can be distinct as the EU umbrella framework allows for diverse and even more ambitious goals under its parameters. Nicole Kobosil from the European Environmental Agency summarizes this as follows:

> The EU Commission has the responsibility of making new legislative proposals. These proposals are then adopted by the Council and the Parliament, and implemented in the Member States by the national authorities. The Commission has the responsibility of overseeing that Member States act in accordance with EU law.
>
> (pers. comm., 1 July 2015)

The NEC Directive (2001/81) focuses on long-term targets and reduction curves with nationally differentiated emission ceilings, thus allowing countries like Denmark to set goals that are much more ambitious than most of the other EU members. The different adoption speeds for individual governments are important indicators of how the power dynamics between more supranational and national forms of governance operate.

Simultaneously, according to Hey, 'national environmental policies have become mainly EU driven' and more careful, as the Sixth EAP focuses on 'cooperative approaches with industry, such as integrated product policies, the wider use of standardisation for environmental policies, voluntary agreements, cooperation with Member States' expert fora' (2005: 27). Two key areas emerge from these arrangements. The first is to do with a constant shifting of responsibility between the EU and the individual states on policy and accountability. There is a pressing need for unilateral environmental protocols to account for different trading standards and policy variations, for example. Thus, the work of organizations like the European Environmental Agency (EEA) is central, as it 'produces reports on the state and trends of the environment in Europe in its attempt to improve the environment and move towards sustainability in Europe' (EEA 2015). The second area of development comes from organizations such as NGOs and consultancies providing specialist knowledge and technical ability in these areas to smooth over potential differences. We have seen the roles of these facilitators emerge as key nodes of the sustainable media network, including the work of consultancies like the Responsible Media Forum as well as a host of European Film Institutes developing environmental sustainability policies (more on these later).

Policy discrepancies

This discussion of general patterns in EU environmental policy has highlighted several key strategies that ensure sustainability is integrated into the fabric of the regulatory regime of each member state and the region as a whole. But when it comes to sustainable media, there simply is no comprehensive framework in place, nor are there clear motivations for establishing policy on the domestic level. Significantly, these complications echo other areas in consolidating patterns of general media policy in the EU. Without delving too far into the particularities of these policies, legislating for integrationist policy for the EU market as a whole has tended to be difficult in reality. Klimkiewicz suggests, for example, that 'media pluralism remains an important value and objective of European media policy, but there has been a lack of consensus and will to apply a harmonised regulatory mechanism to protect media diversity at EU level' (2010: 15). Here, areas of concern emerge from not only the necessity of respecting domestic priorities and regulations, but also from different paces of technological transformation and integration among the regional media institutions: 'Highly complex media policy patterns and trends of functional convergence, institutional interdependence and complementarity, combined with a restrictive mandate based on the EC treaty have conditioned relatively asymmetric policy in media and communication' (ibid.: 14). These discussions also play out in debates over the EU single digital market, where the challenge is persistently one to do with establishing policies that work intergovernmentally and domestically and meet both political and economic needs. To summarize, the establishment of environmental media regulation in the EU faces two major obstacles. One is to do with the complexity and diversity of regulatory mechanisms between intergovernmental and domestic laws as they often fail to satisfy all parties or meet the specific socio-environmental conditions of each nation. Another concern is the complications involved in establishing environmental parameters and strategies that would work collaboratively with domestic emissions policies as well as act as clear incentives for the industry.

Identifying responsibility

The lack of clear oversight on legislative responsibility is at the centre of debates on environmental policy, which becomes more apparent if we focus further on social and cultural forms of sustainable development. A key initiative in this regard is the Creative Europe programme which aims to facilitate 'cross-border cooperation projects between cultural and creative organisations within the EU and beyond; networks helping the cultural and creative sectors to operate transnationally and to strengthen their competitiveness' (Creative Europe 2013: 1). Documents published by the venture cover areas from diversity in the industry to ensuring an equal level of access to production and content. The programme includes these strategies as central KPIs:

- Assessing the EU copyright rules.
- Improving SME policy through the Small Business Act.

- Assessing the impact of online distribution on the audiovisual sector.
- Increasing the availability, use, and re-use of data while promoting cultural diversity, in line with the Digital Agenda for Europe.
- Addressing the challenge of convergence between the online and the physical environment in the application of VAT rates. (ibid.: 2)

These are all central strategic concerns for contemporary media CSR programmes and fit with the mandate of organizations like the Responsible Media Forum. Yet, despite this wide focus on responsibility, the programme has not outlined any response to environmental problems. Similar omissions span the regulatory environment of the creative sector, as can be seen with the Creative Europe Desk UK programmes, *Support for the Audiovisual Sector* and *Support for the UK Culture, Creative and Heritage Sector* (European Commission 2017a). Both make no direct mention of environmental concerns besides a few notes on environment as heritage. This is another frequent framing mechanism that quantifies the value of the environment as human cultural capital. It focuses on a thoroughly anthropocentric approach that sees the environment as a repository for cultural memory, that is, for human consumption.

If the environment is addressed at all in these policies, the focus is on the media's communicative abilities. For example, the European Commission's *LIFE – Programme for Environment* 'is the EU's financial instrument supporting environmental, nature conservation and climate action projects throughout the EU' (European Commission 2017b). The project's focus is on areas like air quality, freshwater conservation, climate change and ecostructures. It contains a section on communication, but the areas discussed focus on awareness raising and environmental training – this is about the ability of the communications infrastructure to disseminate information to the public. The incentive also supports sectors including electronics, buildings, chemicals, machinery and food, but there is no mention concerning the production process for the media. It is clear that in the scale of EU priorities, environmental management of the media does not hold much significance as a potential topic for development.

This argument is explicit in a lengthy response from the European Commission's Eco-innovation and Circular Economy unit who suggests:

> I would tend to agree with you that a good number of environmental regulations currently in place have a direct or indirect impact on the media sector, be that through regulating the resource efficiency of specific machines and tools through the Ecodesign and Energy labelling directives, through regulating waste, the use of chemicals, water, etc. These regulations address, however, any type of business conducted in the EU, without making any specific case of the media sector.
>
> The only EU instrument that I could mention specifically that would be applicable to document the environmental performance of the media

industry and its willingness to undertake improvements in its operations would be the EMAS, which is a support tool for companies and other organisations in reporting on, and improving, their environmental performance. The EU Eco-Management and Audit Scheme (EMAS) is a management instrument developed by the European Commission for companies and other organisations to evaluate, report, and improve their environmental performance. EMAS is open to every type of organisation eager to improve its environmental performance. It spans all economic and service sectors and is applicable worldwide.

(Hugo Maria Schally, pers. comm., July 2015)

These comments emphasize that the sector's environmental emissions are not considered sufficient enough to establish bespoke legislation. The lack of targeted voluntary or mandatory obligations makes individual responsibility even more significant in comparison to institutional responses. Strategies focusing on individual motivations for key personnel – as seen in the managerial strategies of BAFTA – are a starting point, but these do not mean much ultimately if they do not translate into institutional support, as has been the case with the BBC's policy on mandatory CO_2 reporting. In an article on the complex modes of human responsibility over climate change, Brad Allenby suggests any response to the impacts of anthropogenic climate change require dialogue on multiple levels:

It does seem appropriate to impose an individual ethical responsibility to ensure that mechanisms are established by which scientific and technical communities, and society at large, can dialog with complex adaptive systems. However, the challenges of maintaining an open-ended and temporally unlimited macroethical dialog requires an institutional rather than an individual host.

(2004: 11)

These approaches emphasize the need for a strong infrastructural network of individual producers and organizations to consolidate sustainability as a strategic interest for the sector.

Thus, the next section will focus on regulators and ministers in the Nordic countries to evaluate the role of potential players in the network as well as address some of the dynamics that complicate establishing sustainable policies for the sector. The process of securing information from these organizations has been very difficult despite the transparency mandates of public organizations. Many organizations did respond to queries on their approach to sustainability and the media, yet in most of these cases, they have had to respond with a polite decline, citing either a lack of knowledge and experience in the field or of not having anything in place for environmental sustainability. This is illustrative of the larger uncertainty concerning environmental sustainability for the media, which continues to remain a marginal

area of interest. To analyse these circumstances, we now trace the power relations and agentic roles of different organizations in this 'unformed' network.

The Nordic Council of Ministers

The Nordic countries are part of the United Nations Framework Convention on Climate Change and the Kyoto Protocol, and are therefore obliged
to inventory and estimate the amount of greenhouse gas emissions produced
in their societies. The Environmental Protection Agencies of each country
collect and compile information and report to the respective governments.
From there, the Nordic Council of Ministers operates as the intergovernmental body facilitating collaboration between the five Nordic countries on
areas of mutual concern. One of the key principles it works to enhance
through its activities is cooperation: 'Our national societies are based on the
same fundamental values, such as democracy, human rights and sustainability' (Nordic Council of Ministers 2014), and thus, the Council acts as a
mediator between the different governments on these values considered
essential parts of the fabric of Nordic societies. Here, the Committee of
Senior Officials for the Environment are responsible for the implementation
of the Nordic Environmental action plan and argue that 'achieving sustainable
development is an ambitious but necessary goal. There is no other alternative: we must improve global welfare and quality of life while conserving
the capacity of the earth to support life in all its diversity' (Nordic Council
of Ministers 2017). To ensure the Nordic countries engage with sustainable
development productively, the Nordic Council has established indicators for
social, political and environmental areas: 'the Nordic indicators for sustainable development show long-term trends in the following focus areas: the
Nordic welfare model, robust ecosystems, climate change, sustainable use of
the earth's resources, as well as education, research and innovation' (ibid.).

Environmental sustainability is thus one of the priority areas for the
Council as it has active strategies for the protection of ecosystems, education
initiatives, climate politics and renewables energies, as well as the development of environmental labels such as the Nordic Swan. In addition, they
host the Committee for a Sustainable Nordic Region that oversees these
strategies. They process cases from enhancing consumer roles in combating
climate change to developing means to meet international climate standards
across different sectors. These are all areas that would apply to some of the
practices of the media industry, but there are no targeted policies aimed at
the sector developed by the Committee. This omission is especially problematic as Nordic cooperation is framed as an essential concern by the Council
due to common models of cultural policy and exchanges of common cultural
experiences as well as a considerable cultural budget.

Of course, one could argue that sustainable practices in the media simply
are not part of the remit here as most of the strategies of the Council target
areas that directly relate to natural resources (fisheries, forests) or on areas

such as urban conservation or the percentage of renewables in total energy consumption. Sustainable media practice does not seem to fall within its mandate, nor that of another potentially responsible body in the region, the Committee for Knowledge and Culture in the Nordic Region. The Committee has a mandate to oversee areas to do with culture, knowledge and education and includes media and film as key priorities. They have published case studies of relevant media areas such as freedom of speech, internet privacy, journalistic standards and the common digital market, but sustainability is predictably absent. While expecting it to appear here could be unrealistic, its absence from both sustainability and the cultural policy of the Nordic Council indicates problems of accountability and responsibility. Thus, we can conclusively state that institutional support for an environmentally sustainable media policy does not exist on a regional level in its current state.

Individual EPAs

Denmark

These patterns are repeated throughout the sector in all the Nordic countries. As argued, the UK network – our positive case study – works well as responsibility is largely predicated on individual motivation and institutional oversight. Arguably, the comparatively unclear goals and uncertainty over the directions of governmental environmental policy in the UK as a whole have meant that institutes like BAFTA have had to take charge of policy for their sector. As we have already established, the Nordic countries have invested substantially in their infrastructural environmental responsibility. The practical implementation of their strict environmental regulations has arguably led to a lack of urgency on developing strategies for the comparatively marginal area of media production. This encapsulates the response provided by the Danish Ministry for the Environment, who suggests that Denmark's advanced policies and leading role on environmental responsibility mean that the society as a whole is responding to these concerns, indicating that any particular concerns from media production are, in one way or another, already addressed. This exonerates, at least in their view, the Danish Ministry for the Environment from having to provide a bespoke sustainable media policy. While governmental strategy emphasizes the role of sustainable development in most areas of contemporary Danish society, there is a clear responsibility deficit at work in these regulatory structures. The Danish Environmental Ministry's response reflects the ambiguity over allocating responsibility:

> The Danish EPA does not currently regulate the media industry specifically and we do not currently have any projects or other forms of cooperation with the media industry in regards to minimizing the environmental footprint from the industry. I am therefore not sure that we

can provide any valuable insights or knowledge to your project. I would advise you to try and contact the Danish Ministry of Culture, as the media industry in general belongs to their field of responsibility.

(Sune Kirkegaard Rotne, pers. comm., 25 March 2015)

The Danish Ministry of Culture offered to undertake an interview and expressed much interest in the theme. However, their take on sustainability is part of the sustainable development agenda where the focus is on socio-political areas. For example, they actively participate in work on the cultural environment, meaning a perspective where nature, if it is featured as part of this project on developing cultural richness in Denmark, is perceived as cultural capital and subordinated to heritage conceptualizations. While using the environment as a cultural resource can be useful, the Ministry tends to approach the debate from a more anthropocentric angle where the environment has value as a resource to enhance social value systems. For example, one of the key prerogatives for the Danish Agency for Culture is 'to increase cooperation among, inter alia, education, teaching, research, the environment and nature, and business development, including architecture and tourism' (Council of Europe 2012: 13). The environment is a priority but this does not translate into exploring the impact of cultural or media production on the environment. Similarly, cultural production in Denmark is very reliant on public support, but auditing and reporting on sustainability considerations are not compulsory for public producers such as Danmarks Radio. For example, the Minister of Culture's the Media Support Project from 2011 explored the need for 'tomorrow's public media support', but focused on digital media, social networking, and optimal conditions for competitiveness and diversity. These are all obvious parts of the sustainable development agenda covered above and thus of a more anthropocentric perspective on cultural policy.

In discussions with both the Ministry for Culture and the Ministry for Environment, several arguments complicate the possibility of introducing sustainable media policy in Denmark. The Danish Ministry for Culture predominantly focuses on the promotion of environmental awareness as part of their operational mandate (pers. comm., 29 April 2015). However, another complication in promoting awareness concerns what is known as the 'arm's length principle', a policy in practice in all the Nordic countries, which prohibits institutions governing or providing support for publicly funded media from influencing content. This is a major part of the ways media impartiality operates but it also limits the extent to which key development areas – such as environmental issues – can be productively supported. While there are general indications of support for environmental content, these are not clearly explained by policy frameworks, and the arm's length principle makes it difficult to emphasize the importance of environmental content outside of their educational value, perhaps.

Simultaneously, the Ministry for the Environment has no particular relationship with the media industry, even in terms of promoting environmental awareness. These ministries were the obvious points of contact in ascertaining potential OPPs for Denmark but as they have no particular focus on sustainable media management, we had to focus on other organizations with a less visible connection with media practice. These included the Ministry for Climate, Energy and Building, but were unable to 'recognize' anyone who would be able to help (Nikolaj Hvillum Ronsbo, pers. comm., 2015) and the Danish Business Authority. The latter run a unit on green transition but do not have any specific policies on the environmental regulation of the media industry (Dorte Vigso, pers. comm. July 2015). The Danish regulatory environment shows a distinct deficit in terms of identifying responsibility and accountability for the media's environmental role.

Trying to trace where the responsibilities for developing environmental sustainability policy lies has been difficult. The regulation of multinational Nordic companies like the publishers Bonnier and Egmont – headquartered at Denmark – would fall under EU directives due to the international scope of their operations. Sustainability directives for domestic entities such as the Danish Film Institute would not be the responsibility of the EPA as these would fall under the mandate of culture. Guidelines for products and energy used by the media industry (beyond WEEE or RoHS) would need to adhere to the same standards as other sectors and would not require new legislative guidance. The recommendations on green public procurement apply to ICT used in all sectors and not only the media industry and would thus not require specific regulations for the technology utilized in production of various forms of media. The same would apply to other technological solutions, such as LED lights, as these would not only be a focus for media regulation. Strategies like the Danish Producer Responsibility scheme requiring companies to report on emissions apply mostly to large industry, including some of the larger media companies. Considering these complications, the absence of media-specific regulation is not entirely surprising.

Sweden

These concerns also extend to the other Nordic countries where extensive environmental policies are in place but rarely apply to the media. Of them all, Sweden has some of the most advanced climate goals in the world. The main legal act regulating environmental policy in Sweden is the Environmental Code (1998), which aims to promote all forms of sustainable development. They have also established a significant Generational Goal that guides policy on the non-toxic environment, clean air, consumption patterns and the use of natural resources. These goals are coordinated by relevant Swedish environmental laws that are influenced by EU environmental policy, including the EU Emission Trading directive (2003/87/EC) and the Emissions Trading Act (2004). These exclude the media sector as its emissions are not considered to be of sufficient scale to qualify for trading initiatives nor can they be monetized easily to play much of a role in any carbon trading schemes. Yet, as with Denmark, some of the regulatory advances made in Sweden could work as a solid basis for media policy. In 2009, the government introduced two bills on an integrated energy and climate policy, which, according to Wiven-Nilsson *et al.* (2013), exceed the general EU goals. The bills have two particularly striking aspirations that differ from those of the EU: a 40 per cent decrease of 1990 levels of greenhouse gas emissions not included in the EU Emissions Trading Scheme by 2020; and carbon neutrality by 2050. At the same time, Sweden continues to implement EU rules on emissions trading and geological storage of carbon dioxide, but it also levies 'extensive taxes' on carbon dioxide and sulphur on fuels, and energy tax on electricity. Clearly, then, there would be scope here to redirect these extensive policies to cover the media industry.

Simultaneously, the lack of urgency on establishing environmental media policy evinced in the arguments of the Danish EPA could also apply in light of the extensive regulations in place in Sweden. The Swedish Environmental Protection Agency was asked to elaborate on the Agency's mandate to work on behalf of the Swedish Government on the following remits:

- Compiling knowledge and documentation to develop our own and others' environmental efforts.
- Helping to develop environmental policy by providing the Government with a sound basis for decisions and by giving an impetus to EU and international efforts.
- Joining in environmental policy implementation by acting in such a way as to ensure compliance with the Swedish Environmental Code and achievement of the national environmental objectives (Naturvårdsverket 2017).

The response to whether any of this was media-related clarifies the position of the Swedish EPA: 'Unfortunately, we do not have any activity that focusses on the media industry as such, and therefore I do not think that an interview with us would give you any new information' (Marcus Carlsson Reich, pers. comm., July 2015). As with Denmark, key authorities in Sweden concretely highlight the ways the responsibility deficit operates in these contexts.

Iceland

As with Denmark and Sweden, Iceland appears at the top of global environmental charts and has a particularly strong infrastructural relationship with renewable resources due to its geological features. This has led to a more pronounced role for the environment in sustainable development. For example, the Ministry for the Environment (2007) suggests that equal rights are not only about multicultural diversity but apply to concerns that exceed the present population and must be secured for generations to come. The long-term implications of the right to land and to a clean environment are foregrounded in ways that sustainable development in most cases fails to explicitly do. Iceland aims to achieve this by increased utilization of renewable energy, and by the use of economic and regulatory mechanisms. The plans include increasing the total of renewable energy resources in the nation's energy budget with the aim of making fossil fuels insignificant within a few decades. Electricity and geothermal heating systems ensure a renewable infrastructure and this will be extended to transport – one of the key areas of carbon expenditure – as soon as it is feasible to do so. The Ministry for the Environment also has other objectives for development, including environmental purchasing, and they encourage the adoption of the Nordic Swan eco-label for domestic firms and institutes, more choice for consumers on eco-labelled goods and services, instructions on the value of sustainable day-to-day consumption, and awareness of environmental construction. Yet, as with the other Nordic countries, these advances in the environmental infrastructure of Iceland have not led to any applicable policy on the media.

Norway

Policy developments in Denmark, Sweden and Iceland emphasize the extent to which the Nordic countries set their own policies on sustainability and the environment. While they often evoke the argument that new regulatory standards, such as those applicable to sustainable media policy, would need to come from the EU, they also lay claim to being green 'enough' not to require a new set of regulations for media. Similar patterns apply to Norway with the Norwegian Environment Agency addressing both short-term and long-term climate effects. Norway is particularly interesting as a case study due to its heavy reliance on the GDP contributions of the petroleum industry. At the same time, using renewables forms a large majority of its energy infrastructure. Close to a 100 per cent of energy consumed by the population is generated by hydroelectric plants and there is a heavy level of investment in the development of alternative energy sources including floating wind turbines. Norway was also one of the first countries to adopt a carbon tax in 1991 in an attempt to slow global warming and to capture carbon dioxide and store it underground (Norwegian Environment Agency 2015).

While there is considerable institutional support for environmental policy, bespoke strategies for 'minor' sectors such as the media are lacking in scope. To overcome any potential complications from this lack of policy, the Norwegian Ministry for the Environment makes a particularly powerful case for the role of infrastructure in greening the media. They suggest that more than 80 per cent of 'domestic greenhouse gas emissions are either covered by the emissions trading scheme, and/or subject to tax on greenhouse gas emission' (2007: 7). In addition, all industries face legal regulations while financial incentives and support for research and innovation act as important policy instruments to reduce greenhouse gas emissions. These apply to the media by default as the supply chain of media production will have to undergo controls set by one of these legislative regimes. Consequently, key areas of its footprint have to be subjugated to the strict environmental standards of all industry. In addition, the Norwegian media industries are affected by several environmental and climate policies:

- The Norwegian-Swedish electricity certificate market will increase renewable power supply by 26.4 TWh in the period 2012–2020.
- Norway aims to strengthen energy requirements in the building code to passive house level in 2015 and nearly zero energy level in 2020.
- Ban on the use of fossil oil for heating in households and for base load in other buildings from 2020.
- The government in the state budget for 2016 will suggest a reduction in the electricity tax for large data processing centres. This will increase the likelihood of such centres in Norway, where the ambient temperature is favourable and power production almost 100 per cent renewable. (Norwegian Ministry for the Environment, 2012: 9)

These infrastructural strategies for the Norwegian economy are only tangentially connected to media production, but they do indicate how environmental standards for society and the economy as a whole play into media policy. Similarly, another environmental concern relevant to the creative industries comes from restrictions on imported paper sourced from 'the sale of timber and timber products that is associated to illegal logging' (Kristian Rasmussen, pers. comm., 25 June 2015). While these show a measure of awareness of the specific requirements of the media industry, the Norwegian Ministry for Climate and Environment has developed its own environmental scheme, Miljøfyrtårn (Eco-Lighthouse), to help both public and private organizations, including events and festivals, to improve their environmental performance. This is a voluntary scheme requiring active interest from participating companies, but disappointingly, despite advances in the arts, the scheme has not received much interest from the media sector. Anna Asgard Despard, of the Eco-Lighthouse Foundation elaborates on these advances and problems with industry:

> Among our certified enterprises there is one newspaper – the *Adresseavisa*, based in Trondheim. There are also a few enterprises within the media industry, although mainly pertaining to PR and communication rather than traditional media in the form of news and broadcasting. Consequently, although our enterprises submit a yearly environmental report, there is not much data concerning the media enterprises in our system.
>
> (pers. comm., 15 July 2016)

Once again, the industry's environmental responsibilities are overlooked due to a lack of concrete participation from industry players.

Policy patterns in Finland

While the integration of sustainable media policy into the domestic regulatory agenda struggles in most of the Nordic countries, the argument for an inevitable infrastructural greening of the media does make sense. Yet, this does not address many of the key questions over sector-specific considerations that are much more complex than can be covered by unilateral infrastructural transformations. To address these issues in depth, we focus on Finland to evaluate the challenges for constructing policy for a sustainable media network. Finland provides a useful case study as many of its regulatory institutes profess an interest in environmental matters and its media sector has developed intriguing advances in policy. Of all the Nordic countries, it appears the closest to the level of the UK with national broadcaster YLE leading the region in the level of development and scope of its policies in addition to the private sector providing international benchmarks in publishing. Yet, many parts of its regulatory environment also showcase the responsibility deficit and problems concerning merging ethics with business. For one thing, the Finnish law on energy requires environmental reporting from organizations whose turnover is more than 50 million euro. This excludes the majority of media companies from consideration as they are best characterized as small or medium-sized in the scope of their operations.

This is just one area of concern that leaves the industry in a liminal position, which requires both institutional support and individual investment. At the institutional level, if a sustainable media policy was to be developed, the Ministry for Culture and Education would be a key OPP. They have a mandate to develop cultural initiatives and ventures that shape the future patterns of the communications and media industry. The Ministry has characterized sustainable development as a key part of its mandate to enhance cultural activities. While strategies such as the Kulttuuripolitiikan strategia 2020 [Cultural Political Strategy 2020] (Opetusministeriö 2009) host a significant role for sustainability, the implications of this strategy for environmental sustainability are far from certain, as most of the activity is based on the anthropocentric areas of the sustainable development agenda. Here, this is predominantly concerned with education and societal welfare, though explicit environmental concerns are mentioned on a few occasions. For example, it covers areas ranging from recycling to stopping bullying, from cutting down on energy to enhancing the work environment. At the same time, 'it partakes in charting threats generated by climate change and instigating necessary actions' (ibid.: 15). Yet it is not clear what form these actions would take.

One of the 'key projects and responsibilities of the Ministry are preparation of a cultural environment strategy and creation of favourable conditions for creative industries' (Opetus- ja kulttuuriministeriö, Ympäristöministeriö 2014: 3). Consultation for the kulttuuriympäristöstrategiaksi, 2014–2020 [the Cultural Environment Strategy] was conducted in collaboration with the

Ministry for the Environment (another key OPP, as we will see), but these strategies were concerned predominantly with providing conditions where arts and culture can grow and compete effectively. This approach characterizes how cultural policy is formulated in Finland, and how its ability to incorporate environmental sustainability is inherently flawed from a conceptual basis. The rhetoric used by the Cultural Environment Strategy is full of obtuse language that indicates a lot but contains relatively little tangible material:

> The cultural environment is a significant cultural, economic, social and ecological resource and facilitator of new activities. The strategy aims to increase understanding of these opportunities.
> The strategy aims to strengthen sustainable development and the ecological, economic, social and cultural values associated with it by using good care and responsible development.
>
> (ibid.: 7)

If these were taken as the types of ordering discourses identified in earlier chapters, the vagueness of the rhetoric positions the cultural environment as an anthropocentric concept. Yet, some of the sections of the earlier Cultural Political Strategy 2020 do provide a more focused perspective on the environment, including sections such as 'Can Development Be Sustained Culturally?' [*Kestääkö kehitys kulttuurisesti?*]. Here, environmental considerations are discussed explicitly, following a narrative that flows from ethical understanding to environmentalist action to more societal forms of sustainable development:

> The values of citizens will undergo change and ethical and ecological questions will be more pronounced targets of societal discourse and politics. Climate change is a central area here. Its impacts must be taken into account in protecting the built and other cultural environments.
>
> (Opetusministeriö 2009: 19)

Environmental concerns are framed as a threat to the anthropocentric cultural environment, which, by extension, needs protecting via appropriate policy. Taken critically, mitigating the consequences of climate change are not positioned as the central concern for the Ministry for Culture and Education (presumably due to other governmental departments focusing their attention on it), but instead its impact on ensuring a society that thrives culturally seems more pressing. Cultural progress is, thus, a priority where 'sustainable development needs global social responsibility. A culturally sustainable development means respect for creativity and cultural diversity' (ibid.: 19). The existence of a culturally thriving society is thus premised on an anthropocentric understanding of socially and environmentally sustainable habits.

It is not surprising that cultural sustainability emerges as the key priority for the Ministry for Culture and Education considering its mandate. Yet, identifying a regulatory organization responsible for sustainable media is difficult as the Ministry for the Environment, for one, urged me to turn to Ministry for Culture and Education to discuss precisely this topic. Neither organization has invested in media-related sustainability, and especially with the Ministry for Culture and Education, arguably, anthropocentric concerns overtake other considerations. This is especially prescient in the framing of the cultural environment as a productive framework to develop new business activities and corporate innovations to facilitate financial and managerial incentives. To illustrate, collaborating with the framework of the 2007–2013 European Social Fund (ESF) programme, the Ministry for Culture and Education financed a national development programme entitled 'National programme for promoting the growth and internationalisation of the entrepreneurial activities in the creative industries'. The aims of this programme were:

- promoting R&D and innovation activities in the creative industries;
- affirming entrepreneurial competence;
- enhancing the competences of producers and managers;
- analysing and affirming knowledge needed to anticipate changes in the operating environment of creative industries (Opetusministeriö 2010: 9).

The emphasis of these strategies resides clearly in competition and the 'professionalization' of the role of creatives in these industries. If effectiveness and competition are key areas for cultural policy, they also contribute to circumstances where sustainable development is divested of its connotations as a means to conceptualize more dynamic environmental practices. According to Mitchell and Kanerva, 'In recent years managerial-technocratic issues have outnumbered ideological and value issues in Finnish cultural policies and also in public debates' (2014: 44). The addition of an environmental dimension to cultural policy would comprise another regulatory layer to an already complex bureaucracy underlying the operations of cultural organizations. Considering the emphasis on efficiency and competitiveness, it is not surprising that an environmental sustainability policy for the media is largely considered the responsibility of general governmental policy on the environment. As a consequence of all of these areas, environmental sustainability as a cultural incentive does not find much foothold in cultural policy discussions.

The situation is not much more different with the case of the Ministry for the Environment. They have no bespoke strategies targeting the media's environmental footprint but include it as part of other societal programmes. One of the key initiatives they have developed is Social Commitment 2050 (*Yhteiskuntasitoumus 2050*). This is a voluntary incentive that aims to encourage organizations to commit to sustainable protocols. It emphasizes three areas: (1) effective acts; (2) a responsible production chain; and (3) generation of awareness over the environment. Yet, only very few media companies have participated in this venture. The Finnish National Broadcasting Company YLE holds a key role as a leader for the media industry, arguably due to its governmental mandate, but others had not even heard of the strategy in 2015, including the large publishing house Alma Media. This, to me, suggests a lack of correlation between the sectors and a sense of marginalization of cultural areas in the strategy of the Ministry for the Environment. In addition, the Ministry for Culture and Education has had no involvement in this programme, enforcing the sense that there is a lack of collaboration between these organizations who are ideally positioned to consolidate a more comprehensive understanding of the media's role in culture and the environment.

As neither of the ministries most obviously connected to sustainable media policy claim responsibility over the development of environmental incentives for the media, the framing of sustainability as a matter of commerce and technology could provide an alternative angle. This would make the Ministry of Economic Affairs and Employment and the Ministry of Transport and Communications potentially responsible. The former oversees competition and encourages growth while the latter is in charge of the communications infrastructure. Neither has a direct focus on overseeing integration of sustainability into specific media production or distribution practices. The Ministry of Economic Affairs and Employment is invested in, for example, facilitating the capability of large-scale media companies like Alma Media to compete, but they have no professed oversight over environmental matters. The Ministry of Transport and Communications may have into account some environmental areas including the energy usage of communication networks and their own operational footprint. But most of their attention is focused on ensuring the communications infrastructure is prepared to aid preparation against the destructive effects of climate change by, for example, facilitating better communications in times of crisis. As this is a ministry focusing on both transport and communications, the majority of their policy touching on environmental strategies focus predictably on emissions from travel. Thus, sustainability initiatives fall into a liminal space between the ministries as the cultural and environmental ministries who may have potential investment in the theme have no real initiatives in place for developing sustainable media policy due to its marginality in relation to their KPIs. The economic and communications ministries do not see it as a relevant problem due to its roots in cultural creativity or their prioritization of areas with a much more visible, as well as public, footprint, such as the use of various energies in transportation.

The Principality Decision

The responsibility for implementing sustainable policy is not the sole property of various ministries, of course. The State Council's Principality Decision (*valtioneuvoston periaatepäätös*) is a mechanism designed to provide guidelines and advice to the highest echelons of governance. These are considered political directions for the government rather than binding legislation for citizens. The 2012 Principality Decision on Social and Corporate Responsibility is of relevance to the potential sustainable media network. It focuses in part on sustainable consumption and production and suggests that ecological questions have a substantial role in tax design and industrial production. Unsurprisingly, environmental production standards identified here concern large industrial sections of considerable importance for the national economy. These include areas like mining or energy with the ecological role of media production, unsurprisingly, not mentioned.

Yet, several areas are of relevance to our inquiry as, for example, the Principality Decision on 'the advancement of new and sustainable environment and energy solutions in public procurement' (Valtioneuvosto 2013) means that the state and municipalities have to take green technology into account in all their procurements. These include the construction of new cleantech areas and enhanced use of the environmental labels developed by the EU and the Nordic countries. In addition, companies purchasing products and services can ask their supply chain to provide them with information on their environmental programmes, which comprises a strategy to encourage such companies to initiate them if they do not already have them. This is all rationalized by suggesting, first of all, that the use of natural resources has expanded significantly and, second, that the financial investment in cleantech and renewables is expected to increase exponentially. The benefits of sustainability measures come from three areas: the economic, the organizational, and the environmental. Economic concerns emphasize that sustainable means are not necessarily more expensive than traditional sources of energy, especially if we take into account the lifecycle of renewables and their potential to anticipate future transformations. Organizational benefits come from both governmental investment in green areas and their ability to attract private investment. Finally, environmental areas can be used to create new markets by innovating and diversifying the integration of resources into areas such as travel and building construction.

It is clear from these strategies that the development of environmental media policy is not productively integrated with governmental responsibility. Thus, it is not surprising that both the Ministry for Culture and Education and the Ministry for the Environment suggest that responsibility for environmental practice in the media may reside with Tekes (an organization focusing on financing of technology and innovation) and Finpro (supporting companies in exporting environmental technology). Their roles as mediators between business and the creative sector make them particularly powerful potential OPPs for integrating sustainability into the cultural sector. Both have a professed interest in investing in environmental innovations to advance competitiveness in the Finnish industries. For example, one of the mandates for Tekes is to provide funds for technological innovations that advance the sustainable uses of energy and natural resources and their responsible application to a range of industrial settings. Finpro, meanwhile, emphasizes different areas such as Finnish cleantech and public procurement innovations and works to facilitate the impact of environmental innovations. The work of Tekes on developing new technologies and Finpro of consolidating best practice as an industrial norm are some of the key strategies a functioning network requires as both organizations highlight some of the ways infrastructural considerations could be used to enable the consolidation of environmentalist practices in the media sector. Similarly, procurement of green technologies by public organizations is a central operational strategy to ensure that the infrastructure of companies producing media fulfils sustainability standards. Both organizations would have the means to take environmental media strategies on board and both are expected to do so by the ministries in charge of key areas overlapping sustainable media. Yet, neither have, at least at the time of writing, moved to initiate any specific policies for the media sector.

While none of these organizations have developed policy for environmental media, their approaches hold substantial potential as they gesture to a merger of ethics and commerce that sees tackling climate change as good for both business and the planet (providing a form of ethics collapsed into self-interest which reflects the approach taken by other climate conscious organizations as the Institute for Public Policy Research document *Warm Words* (IPPR 2007) suggests). In many ways, similar to the IPPR, they perceive climate change as 'a cost, as opposed to constructing measures that counter climate change as a cost' (ibid,: 21). Here, tackling environmental costs is seen as a value in its own right, which provides a means of mitigating inevitable costs on core business practices instead of focusing on the costs of implementing new policy. The rhetoric uses business logic in a way that frames sustainable activities as a common-sense cost-saving exercise, highlighting the ways corporate strategies benefit from the integration of environmental sustainability into organizational management.

This approach leads to another potentially significant OPP for the industry to integrate sustainable measures – the Finnish Business and Society (FIBS) network. This is a leading consultancy coordinating responsible business strategies for Finnish companies and acts as a key instigator of awareness in the industry community by, for example, hosting a forum for sustainable development. Part of its role is to oversee that fossil fuels are being eliminated to abide with general climate goals, but, as is the case across the Finnish CSR spectrum, FIBS does not have any particular strategies for media organizations. This is especially significant as FIBS would be an essential part of a sustainable media network, but as they have not considered the footprint of media production in any particular detail, they reflect the slow level of development in the Finnish sustainable media management network. While many parties share an interest in CSR and environmental sustainability, their efforts have not congealed around the media due to a lack of regulatory oversight and absence of coordination from the sector. In addition to this responsibility deficit, the scope of emissions and the lack of unified standards to measure and counter these emissions have contributed to the marginal role environmental media policy has in this context.

A further concern with allocating responsibility lies in deciphering what exactly comprises the responsibility of a media organization and what areas would (and could) be legislated under sustainability regulation. To identify different types of emissions and managing their reporting, emissions reporting uses 'scope' classifications to indicate the different types of emissions generated by industrial activities. Scope 1 emissions consist of direct emissions from production activities in areas directly owned by the company. These include areas such as fuel used by company vehicles on transportation of materials or energy consumed in production activities. For emissions from media companies, these would include areas like transporting crew or gear between sets or the use of specific lighting arrangements in the studio. Scope 2 emissions are based on infrastructural materialities which include the purchase of energy for most areas of a company's operations including fuel and heating. These emissions are part of the immediate responsibility of a media company and include elements such as the use of renewable energy sources at company offices. They tend to be negotiated in line with company policy and provide one of the key steps for a media company to claim they have attained sustainable goals.

Scope 3 emissions are the result of the activities of the company but are outside of their immediate means of control (as is the case with production emissions and electricity purchase, for example). These include the following: purchased goods and services, business travel, employee commuting, waste disposal, use of produced products, transportation and distribution (up- and downstream), investments in other companies, environmental performance of leased assets, and external franchises. For media, these are to do with the distribution of content, the carbon footprint of materials used for sets or other production materials, staff transport to the office and to meetings, recycling of used materials after they have been processed by the company, the impact of minerals, metals and such for media devices and the processes for dismantling and recycling them. Additionally, they also include the form of media on which users consume content and their means of consumption (i.e. paper, film, DVDs, digital). These are all areas that are outside the direct influence of the media company and involve a wide range of stakeholders from the upstream supply chain and to the downstream consumer patterns. They are thus very difficult to measure, especially in terms of supply or value chain analysis that could lead to tracing an accurate footprint.

The UK has made reporting of Scope 1 and 2 GHG emissions mandatory for all quoted companies under the Companies Act of 2006 (Strategic and Directors' Report) Regulations 2013. Scope 3 emissions are not included as mandatory as the range of areas is still too complex to assess appropriately. A study by the VTT Technical Research Centre of Finland and KTH Royal

Institute of Technology in Sweden (Hohenthal *et al.* 2012) estimates that the majority of emissions are generated by consumer choices in accessing said content (going as high as 87 per cent of total emissions for online publications). At present, the majority of media companies operating in the UK do not report their footprint as their operations tend to be too small in scope and the companies are not listed on the stock market. These requirements make it difficult to impose regulation on the industry. Further concerns include the supply chain outside of the direct control or ownership of media companies. Only very large companies such as service providers Google or Facebook or manufacturers Apple or Samsung face circumstances where the carbon intensity of their core competencies is high enough to make Scope 1 and 2 emissions a strategic or material priority. For most of the media, Scope 1 and 2 emissions tend to be relatively minor. Yet, Scope 3 is where future developments will need to take place.

Even as Scope 3 emissions continue to pose problems for a variety of sectors, Nordic environmental regulation is catching up on the scope emissions of their industries. A report produced by the sustainable development consultancy CDP (2017) focuses on the latest developments (for 2016) in the Nordic countries based on emissions data from the leading high emissions sectors from the region. The report shows that the response to climate mitigation has been largely positive in the Nordic countries. However, while the study provides data on the ICT sector, media companies only feature in the Consumer Discretionary category, which here comprises a sector that provides non-essential services. While major publishing houses Alma Media and Schibsted are featured under this category, it is clear that in relation to total emissions in Nordic societies, the media is not seen as a relevant sector as it is classified alongside retail and consumer goods service providers, all sectors with limited emissions in relation to heavy industry. One of the reasons for this oversight may be the lack of comprehensive data for Scope 3 emissions from media producers. Some 80 per cent of the companies that participated in the study provided data on Scope 1 and 2 emissions, and 62 per cent on two or more Scope 3 emissions. At the same time the report is very critical of these companies for having no real pragmatic plans to measure their Scope 3 emissions as they have failed 'to grasp the importance of reliable and complete data for decision makers' (ibid.: 15). This indicates that there is still a lot to accomplish in terms of comprehending the role of carbon emissions for high emissions sectors like electricity or utilities, let alone the media. Yet, if we include the potential impact of Scope 3 emissions for digital media, the equation may change, especially if we consider the huge resource consumption and emissions of companies like Google and Facebook. Considering Greenpeace estimates that the ICT sector contributes 3 per cent to global CO_2 emissions, reorienting the focus on these areas of the media's footprint may change the significance it holds for regulators. Yet, as Scope 3 emissions remain a grey area, significant regulatory or financial attention on the media's footprint continues to be similarly elusive.

Conclusion

The Nordic countries have comprehensive environmental regulation in place, yet this does not apply to the media in any binding way. The existing initiatives tend to be minimal and largely voluntary (or formulated by public institutions such as the national broadcasting services – more on this below). One of the arguments explaining this is that the Nordic countries have much stricter existing environmental regulation frameworks than the UK, for example. These include comprehensive carbon targets that apply to all sections of the societies and enforce measures on most industrial sectors. Such extensive regulatory frameworks suggest that a specific media policy focusing on curtailing the emissions of the media is not necessary as the Nordic countries are in many ways green as it is, and the imposition of similar measures as with BAFTA, for example, would create unnecessary obstacles for the industry and not result in substantial environmental consequences.

This argument contributes to what I have identified as the responsibility deficit of the Nordic regulatory environment. Sustainable media practice could fall within the mandate of several fields of governance. The allocation of responsibility to cultural or environmental ministries does not seem to result in any real consequences as neither considers themselves to have direct responsibility over both the environment and the media. Thus, designating a particular ministry or a singular body to oversee the area would face challenges of jurisdiction that would need to involve multiple stakeholders and interests. Coordinating management of the media industries alone would need to involve concerns such as labour laws, human resources, ICT, energy infrastructure, and protection of the environment, and these would have to be evaluated on an equal level for any regulatory frameworks to operate effectively. This level of oversight would be difficult to allocate to a governmental ministry in ways that would not overstep its legislative powers and fall foul of other actors like NGOs. Due to the governmental delegation of responsibility, departments often seem focused on particular aspects of environmental management instead of considering the bigger picture – i.e. cultural departments tend to look after value systems to do with cultural 'sustainability' whereas environmental ministries address material notions of conservation and waste. Of course, EPAs do tend to promote cultural values including communication about key environmental issues, but much more could be done in terms of addressing the overlap between the categories of culture and the environment (especially as pertains to the role of the media).

The complexity of assigning responsibility should not lead to the avoidance of the entanglement of cultural and environmental questions in governance. The current lack of intervention can, in some ways, be attributed to the arm's length principle that affords creative organizations a large degree of autonomy. Yet, this principle would not restrict oversight of environmental management as such strategies could be considered tactical interventions much like other media policies, including digitalization of cinemas (a particularly comprehensive and successful policy). Simultaneously, there is a very fine line between infrastructure and imposing constraints on the freedom of media producers. The imposition of mandatory environmental reporting is still considered problematic as FIBS, for example, suggests that this would contradict the principles of how Finland as a society operates. Due to the strong tradition of media impartiality and freedom, any imposition of regulation, even if this is to do with production practice, is met with suspicion. Coupled with the consequences of the responsibility deficit, this translates to a lack of key OPPs that would, in all instances, coordinate the inclusion of stakeholders into the network. As we have seen, such OPPs are essential in motivating participants – both in the industry and in the supply chain – to take part in consolidating sustainability strategies for the media, and as there are few organizations championing such areas specifically for the media sector in the Nordic countries, this leaves a vacuum in the centre of the network.

These complications provide intriguing reflections of what Haluza-DeLay suggests as a lack of environmental awareness in contemporary society as a whole, which does not meet, as she puts it, the qualifications of an 'ecological habitus' (2007), that is, circumstances where environmental values are generally taken as a behavioural norm. In the environmental management of Nordic media, the material realities of the media are not prioritized sufficiently to make this network attain the type of ecological habitus discussed above – environmental questions are not of sufficient importance for governing bodies who oversee the sector to establish new protocols to normalize environmental sustainability in the media. The transferable uses of environmental sustainability dilute its constitution and contribute to the responsibility deficit explicit in much of the environmental media policy discussions in the region. In the more successful networks, such as the UK film and television network, self-regulation and oversight by cultural organizations ensure that environmental media management is integrated into the core competencies of the sector (though it is, once more, worth emphasizing that this is directly in response to the lack of appropriate environmental regulation at state level, protocols that, interestingly, are much more extensively coordinated on the intergovernmental EU level). The Nordic regulatory network approaches environmental sustainability largely as an infrastructural necessity covered by general environmental regulation. Consequently, sustainability tends to be perceived as only another management strategy complicating the daily organization of the sector. Hence the specific emission types and processes that media production generates are not addressed through organizational management or policy. While regulatory imperatives are lacking, strategies dealing with the emissions of different media have been developed within the industry, but approaches to emissions vary, depending on the type of media production prioritized by specific organizations. As we will see, broadcasting prioritizes different areas from film, which again, generates different production emissions from television production, all which are very different from the publishing industry.

Bibliography

Allenby, B.R. 2004. 'Engineering and ethics for an anthropogenic planet', in National Academy of Engineering (ed.) *Emerging Technologies and Ethical Issues in Engineering.* Washington, DC: National Academy Press, pp. 9–28.

Browning, Christopher S. 2007. 'Branding Nordicity: models, identity and the decline of exceptionalism', *Cooperation and Conflict*, 42(1): 27–51.

CDP. 2017. *CDP Climate Change Report.* Available at: www.cdp.net/en/climate

Council of Europe. 2012. 'Denmark: competence, decision-making and administration', Available at: www.culturalpolicies.net/web/denmark.php?aid=32&language=de&PHPSESSID=9l05hv3fsort4jj11dql50en63

Creative Europe. 2013. *The European Union Programme for the Creative and Cultural Sectors*, Brussels: European Union.

European Environmental Agency. 2015. *Communication, Environment, Behaviour. EEA Report 13/201.* Brussels: European Environmental Agency.

European Commission2017a. 'Representation in the United Kingdom'. Available at: https://ec.europa.eu/unitedkingdom/tags/media_en (accessed 11 November 2017).

European Commission. 2017b. 'Life Programme'. Available at: http://ec.europa.eu/environment/life/ (accessed 11 November 2017).

Greening Film. 2017. 'About Greening Film'. Available at: www.greeningfilm.com/about (accessed 11 November 2017).

Haluza-DeLay, R. 2007. 'A theory of practice for social movements: environmentalism and ecological habitus', *Mobilization*, 13: 205–218.

Hey, Christian. 2005. *EU Environmental Policies: A Short History of the Policy Strategies. Environmental Policy Handbook.* Brussels: European Commission.

Hohenthal, C., Moberg, Å., Arushanyan, Y., Ovaskainen, M., Nors, M. and Koskimäki, A. 2012. *Environmental Performance of Alma Media's Online and Print Products.* Helsinki: VTT/Stockholm: KHT.

Ingebritsen, Christine. 2012. 'Ecological institutionalism: Scandinavia and the greening of global capitalism'. *Scandinavian Studies* 84(1): 87–97.

IPPR. 2007. *Warm Words: How Are We Telling the Climate Story and Can We Tell It Better?* London: IPPR.

Jordan, A. and Adelle, C. 2012. *Environmental Policy in the European Union: Contexts, Actors and Policy Dynamics.* London: Routledge.

Kauffmann, C., Tébar Less, C. and Teichmann, D. 2012. *Corporate Greenhouse Gas Emission Reporting: A Stocktaking of Government Schemes.* OECD Working Papers on International Investment, 2012/01, Paris: OECD Publishing.

Klimkiewicz, Beata. 2010. *Media Freedom and Pluralism: Media Policy Challenges in the Enlarged Europe.* Budapest: Central European University Press.

Law, John. 2007. 'Actor Network Theory and material semiotics'. Available at: www.heterogeneities.net/publications/Law2007ANTandMaterialSemiotics.pdf (accessed 11 November 2017).

Ministry for the Environment. 2007. *Iceland's Climate Change Strategy.* Reykjavik: Ministry for the Environment.

Mitchell, Ritva and Kanerva, Anna. 2014. 'Finland'. In *Compendium of Cultural Policies and Trends in Europe.* Brussels: Council of Europe/ERICarts.

Naturvårdsverket. 2017. 'About the Swedish Environmental Protection Agency'. Available at: www.swedishepa.se/About-us/ (accessed 11 November 2017).

Nordic Co-operation. 2013. 'Nordic Sustainable Development Indicators'. Available at: www.norden.org/sdindicators (accessed 11 November 2017).

Nordic Council of Ministers. 2014. 'The Nordic Region – Together We Are Stronger'. Available at: www.norden.org/en/nordic-council-of-ministers/ministers-for-co-opera tion-mr-sam/declarations/the-nordic-region-together-we-are-stronger/ (accessed 11 November 2017).

Nordic Council of Ministers. 2017. 'Indicators for Sustainable Development'. Available at: www.norden.org/en/nordic-council-of-ministers/ministers-for-co-operation-m r-sam/sustainable-development/indicators-for-sustainable-development-1/ (accessed 11 November 2017).

Norwegian Environment Agency. 2015. *Climate Mitigation Measures and Emission Trajectories Up to 2030.* Oslo: Norwegian Environment Agency.

Norwegian Ministry for the Environment. 2007. *Norwegian Climate Policy.* Oslo: Storting.

Norwegian Ministry for the Environment. 2012. *Norwegian Climate Policy.* Oslo: Norwegian Ministry for the Environment.

OECD. 2011. *Invention and Transfer of Environmental Technologies.* OECD Studies on Environmental Innovation. Paris: OECD Publishing.

Opetus- ja kulttuuriministeriö, Ympäristöministeriö. 2014. *Työryhmän esitys kulttuur-iympäristöstrategiaksi, 2014–2020.* Helsinki: Opetus- ja kulttuuriministeriö.

Opetusministeriö. 2009. *Kulttuuripolitiikan strategia 2020.* Helsinki: Opetusministeriö.

Opetusministeriö. 2010. *Luovien alojen yritystoiminnan kasvun ja kansainvälistymisen ESR-kehittämisohjelma 2007–2013.* Helsinki: Opetusministeriö.

Valtioneuvosto. 2013. *Valtioneuvoston periaatepäätös kestävien ympäristö- ja energiar-atkaisujen (cleantech-ratkaisut) edistämisessä julkisissa hankinnoissa.* Helsinki: Finnish Government.

Van Holten, Saskia and Van Rijswik, M. 2014. 'Consequences of a governance approach in European Environmental Directives for effectiveness, flexibility and legitimacy', in M. Peeters and M. Uylenburg (eds) *EU Environmental Legislation: Legal Perspectives on Regulatory Strategies.* Cheltenham: Edward Elgar, pp. 13–47.

Wiven-Nilsson, Tore, Warmsby, Michael, Edqvist, Madeleine and Nilsson, Sofia. 2013. *Energy Law in Sweden.* London: Wolters Kluwer.

World Commission on Environment and Development. 1987. *Our Common Future* (the Brundtland Commission). Oxford: Oxford University Press.

7 The media in the Nordic countries
Broadcasting

Introduction

The development of sustainable media management in the Nordic countries is premised on a network consisting of intergovernmental and domestic regulators and institutions. Yet, fallacies such as the responsibility deficit and the transferability of the ideological connotations of environmentalism – at times, about ethics, at others about the economy – prohibit this level of oversight from working in meaningful ways. Yet, the incentives that exist to curtail the environmental consequences of heavy industries indicate an infrastructural level of commitment that also applies to the media sector, though often to a different extent than with heavy industry. Considerations to note include the scale of the organizations, the revenue they generate, the types of resources their core competencies require and the organizational infrastructure in which they operate. These translate into a different set of legal and economic requirements on multinational companies like Shell or BP than they do on relatively small-scale Nordic media companies, who lack the capital investment but also whose operational scale exempts them from some of the more intense regulatory frameworks overseeing oil production, for example.

The media environment of the Nordic countries gives rise to the particular manifestations of environmental policy in these contexts. These are small media cultures with a limited range of resources and audiences to support production. Such limitations have a considerable impact on the infrastructures of the industry as the limited audience sizes complicate generating a financial profit from most media activity. The methods to counter small audience sizes vary by sector, with government financing and support existing as an essential mechanism for the majority of the media sector, leading to considerable regulatory oversight. Hence, the motivation to self-regulate is comparatively limited and we see few cultural organizations assuming leading roles at the level of BAFTA, for example. Nordic film and television production is part of a heavily subsidized framework led by domestic film institutes and public broadcasters without which these sectors would not be able to exist. The press and the publishing industry tend to be dominated by companies engaging in a range of print and digital publishing. Consortiums like Alma Media and Sanoma in Finland, or Schibsted and Egmont in Sweden, dominate the markets due to their ability to reach a range of readers through popular tabloids and more specialist magazines, in addition to a host of digital services covering areas like classified ads and data provision platforms. Public broadcasting, of course, is financed by tax money and thus operates under the auspices of ministerial supervision, though with a large degree of autonomy for the broadcasting corporations. In the next couple of chapters, I will focus on these three forms of media production to explore the relationship between the organizing presence of policy infrastructures and the practicalities of sustainable media management.

This chapter will focus on the environmental strategies developed by public broadcasters in the Nordic countries. As is the case with the UK, the responsibility of being a public service broadcaster has implications for the development and implementation of environmental sustainability policies. First, these companies are required to act in the best interests of the public and operate with clear corporate social responsibility (CSR) mandates. Sustainability plays a key part in these strategies by mostly emphasizing notions such as freedom of speech, responsibility, impartiality and diversity. Sustainability priorities are often influenced by the policies set for the cultural sector as a whole, with many of the policies outlined by cultural or environmental ministries falling into what I have called the responsibility deficit for the media. Thus, individual broadcasting companies are required to develop their own priorities and strategies within the mandates outlined by both international and domestic regulation. Much as was the case with the UK and the BBC, some of these companies have developed extensive sustainability strategies while others struggle to incorporate them into corporate policy. Accordingly, we investigate the extent to which the policies of these organizations constitute eco- and anthropocentric forms of management and how they conform to general environmental sustainability patterns in these countries.

This chapter will focus on the Norwegian Broadcasting Company (NRK), Danmarks Radio (DR), the Finnish National Broadcasting Corporation (YLE) and Sveriges Television (SVT) and the ways in which they integrate their public service mandate with environmental activity. It is worth noting that we do not focus so much on the 'materialist ecology' (Maxwell and Miller 2012) of broadcasting (i.e. focusing on the particular technical processes in practice and under development in these companies). Instead the emphasis is largely on policy. This focus provides a means to evaluate the power balances between eco- and anthropocentric perspectives in the planning and implementation of environmental sustainability into the sector. In addition to analysing the strategies of public broadcasters, I evaluate the ways these organizations communicate about their sustainability responsibilities and how these form potentially dynamic networked connections for integrating participants from the supply and value chain. Finally, public broadcasting companies are required to be transparent about their communications and the processes of production as well as report on their CSR operations. These areas make environmental sustainability a potentially contentious issue especially if these strategies are not fully developed, as we will see with our case study of Sweden. Before exploring these complications further, I will set the scene by outlining key strategies implemented by the Norwegian and Danish broadcasters.

NRK: an infrastructural organizational management deficit?

The Norwegian Broadcasting Company (Norsk Rikskringkastning, NRK) maintains independence from the state but is publicly financed through a licence fee. As with the other Nordic broadcasting companies, this relationship sets both expectations and limitations on environmental sustainability within the organization. Accordingly, NRK has to abide by general environmental strategies set by the government and act on both environmental and societal sustainability. To indicate how they meet these goals, NRK has published a long-term strategy for 2016–2021 which contains considerable material on environmental sustainability. The incentive to publish these materials comes from the need of PBS companies for transparency as 'disclosure is a core value. That means we should be open about everything we can and should be open about' (NRK 2015a: 8). These disclosures are part of a government strategy to have all companies report on social responsibility irrespective of their status as a private or a public company. Companies like NRK are expected to work systematically with these areas and act as a leader in the field. Thus, the company must implement environmental reporting on their activities to meet the main climate goals of the government. These goals apply to all Scope emissions and concern both internal organizational management and evaluating emissions from their supply chain.

While NRK has sought to fulfil its mandate on general emissions, these strategies do not concern the sector-specific side of their emissions – such as the variations in material uses and emissions from the three sectors of our focus: film, publishing and broadcasting. The following chapters will outline these in depth, especially as it becomes apparent that the responsibility deficit concerning government oversight also applies, in principle, to how the media industry often approaches the complexity of its own emissions. Here, most of the focus of the industry is on infrastructural concerns which tends to overlook key emissions from production flows and service procurements. To illustrate, NRK (2016) states that they have paid special attention to discussing environmental sustainability in terms of their external environment, which concerns the harvesting of resources to run their operations, including contractors for building construction and maintenance and the incorporation of electricity and energy. Their buildings at Marienlyst, Bergen and Trondheim are constructed to maintain high environmental standards by improving lighting and heating control integration as well as the use of energy-efficient glass for windows. Thus, buildings must be connected to district heating, allowing them to work with the standards established for the society at large.

Furthermore, 'NRK emphasizes environmental impacts in connection with all purchases according to the "Law on public procurement"' (NRK 2015b: 197), which includes essential media production technology, but also a wide variety of equipment for office protocols that are not integrally connected to the core competencies of media companies. NRK states that the completion of energy conservation projects has led to reductions in emissions, which are due to the choice of energy-efficient solutions in conjunction with cooling and district heating, efficiency of ventilation and the choice of modern energy-efficient equipment. Accordingly, oil is only used to power supplies and as a reserve in case of failure of the district heating supply. While much of the GDP of Norway is based on the petro-economy, NRK disassociates itself from its negative associations, thus connecting its operations to the government's mandate to go carbon-neutral by 2030. While these are essential aspects when calculating NRK's footprint, they do not tell us much about how NRK as a media company approaches its footprint.

NRK's strategies on infrastructural emissions management extend to human resources, which focus on ensuring staff compliance with regulations and the internal promotion of corporate values (NRK 2014). The strategies include a list of areas for improving the work environment and general energy efficiency of daily operations. The aim of this is to align the management of staff with organizational goals and integrate them as part of the wider environmental compliance of the organization. Such management strategies are an essential part of normalizing environmental behaviour in the organization by providing guidance for procedures such as recycling paper and food waste, turning off equipment and lights, storing hazardous waste in prescribed containers, to arrangements for obsolete equipment disposal. These extend to NRK leading the sector on establishing protocols on environmental activities including requirements on suppliers and contractors to recycle the waste they generate. NRK has constructed a bespoke facility for recycling in Marienlyst and made agreements with suppliers on returning UPS (Uninterruptable Power Supplies) batteries.

Yet, while they may provide a leading example for the media sector, most of this is in areas where recycled materials are not an outcome of their public broadcasting activities but general operations in areas such as paper use. While some of these advances could be considered media-specific strategies as UPS batteries are essential for broadcasting on location, for example, the main considerations of policy are closer to the daily practice of a large business organization rather than anything that instigates specific strategies for the media sector. As discussed in the above chapter, the Eco-Lighthouse project by the Norwegian Environmental Ministry would be an ideal obligatory passage point (OPP) for the Norwegian media network. Yet, it has had little interaction with or uptake from the media, including the NRK.

The benefits of an organization like the Eco Lighthouse coordinating the environmental sustainability efforts of the media in collaboration with a central organization like the NRK would be to ensure that specific emissions from media production, instead of more general infrastructural concerns, are addressed. Yet, despite the comparably vague strategies of this environment policy, the activities of NRK are grounded on infrastructure which now flows through the supply chain into production practice. Thus, NRK's strategies may not be particularly revealing for the media sector, but they do indicate some of the ways in which the industry is inevitably greened due to these infrastructural strategies.

DR: environmental strategies for a media business

NRK's approach to sustainability is in many ways typical of the sector. The translation of government policy into environmental practice for the media ensures that its infrastructural footprint is accounted for but also that its bespoke needs and emissions are not fully considered. Danish Radio (DR) also takes an infrastructural approach to its environmental management strategies. Annual reports (DR 2014; 2015) outline a range of sustainability initiatives from ethics, digital education, diversity, gender as well as the environment. The documents go through different types of strategic planning which include sustainability and emphasis on a continuous focus on reducing the environmental impact of the organization. The rhetoric in the 2015 report is focused predominantly on easy CSR and PR strategies, avoiding hard or complex data, much like the case of NRK. The discussion explains substantial reductions (from 2011–2015, a reduction of total electricity consumption by 13.4 per cent; 2014–2015, electricity reduced by 1100 MWh) and uses a range of comparisons: 'the reduction of DR's electricity consumption 2014–2015 corresponds to the annual power consumption of 220 single-family houses, resulting in a cut of 550 tonnes of CO_2' (DR 2015: 25). The rhetoric thus positions DR as a responsible operator with a clear focus on the 'big picture' of adopting environmental strategies – much as the Danish Environment Protection Agency had suggested.

DR's strategies for achieving environmental sustainability touch on some of the key areas of the responsibility deficit. There is very little discussion of the kind of work that could align environmental and cultural governance. Instead, the accounting of CO_2 emissions is based on the constants set by the CO_2 calculator of the consultancy Climate Compass (Klima Kompasset). This is a venture led by the Confederation of Danish Industries and the Danish Business Authority under the Danish Ministry for Business and Growth. They operate as part of a larger set of strategies for the Danish government's business climate strategy and work under the supervision of the Minister for Economic and Business Affairs. The positioning of environmental sustainability as part of business operations is standard practice, of course, but the ways these are framed as innovative advances in technology and corporate social responsibility defines them as generalized strategies, despite incidental notes on targeting the core competencies of each of the media sectors. The fact that environmental sustainability is perceived as an operational matter for a business organization indicates the ways it is integrated into the fabric of the media industries. It is thus not a cultural matter so much as a concern to do with the media business. Thus, when it comes to small businesses, which media organizations in the Nordic context mostly are:

> a climate strategy sets out the measures your company intends to take to reduce its CO_2 emissions, some of them being a 'turn-off-the-lights' campaign, purchases of company bicycles and equipment for video conferences as alternatives to transport and investment in energy-efficient machinery and electronics.

> (ibid.: 22)

The infrastructural concerns take precedence over core competencies, signifying that the media is viewed as just another commercial business venture.

As businesses face increasing requirements to report on emissions, for example, on principles established under the EU's directives for waste and facility management, so DR, as a comparatively large-scale media business, has to comply with these demands. DR's environmental and climate accounting efforts for 2015 include CO_2 contributions from direct and indirect energy and fuel consumption in DR buildings around the country as well as from waste disposal. The mechanisms are thorough and take into account the energy losses of electricity and district heating companies during transmission and the employees' transportation during work using trains, rented cars and buses – what would be classified as scope 3 emissions. Thus, both the costs and emissions of staff and infrastructure are addressed in depth. These parts of the document go much more into detail than NRK's reports and include precise calculations of the consumption patterns of the company. One of the key areas includes waste, of which 47 per cent goes to recycling, 47 per cent is incinerated and the rest to landfills. DR emphasizes that waste management is based on employees sorting material into appropriate containers, including processing food waste in canteens. While waste has been reduced by 36.4 per cent, the management of staff behaviour indicates that there needs to be both voluntary action as well as coordination of communications from above for these to take hold in the organization (ibid.: 25).

While DR tends to be very transparent in presenting measurements of its main environmental impacts, the data are presented in ways that leave ambiguities and rhetorical omissions in place. In discussing the organization's consumption of energy, the document identifies heat as a significant factor in the rise of electricity consumption by 2.2 per cent in 2015 at 13,908 MWh. But interestingly, the number of hot and cold days are identified as having an impact on heat consumption in the buildings. Providing a comparison of heat consumption from 2014 and 2015 adjusted for the temperature of the days in question changes the correlation presented and negates the implications of the rise in electricity consumption. The suggestions cast doubt over the veracity of the accounting measurements used for the calculations as they suggest 'a degree day corrected comparison indicates a decrease of 3.8% from 2014 to 2015' (ibid.: 25), thus resulting in the suggestion that DR has actually increased its environmental performance. Another area where seasonal or weather-based considerations influence the material impact – or more appropriately the presentation of the material impact of DR – is to do with water. DR suggests that they consumed a total of 32,081 m^3 of water in 2015, an increase of 4.7 per cent compared to 2014. To limit water consumption, the maintenance of DR has developed a rainwater harvesting system used for flushing toilets. The system has reduced water consumption by 950 m^3 in 2015, yet the total of recycled water has decreased from the 1400 m^3 recycled in 2014. They credit a large part of the reason for the reduction in the recycled total to a decrease in rainfall in 2015 compared to 2014. Such arguments suggest that weather factors influence CO_2 calculations and contribute to DR meeting targets despite increases in actual consumption.

Furthermore, the calculation of the emissions leaves room for interpretation. One of the caveats in these arguments is that DR calculates its CO_2 and climate accounts at the beginning of the year. The calculations for 2015 are based on power and heating companies' emission data for 2014. Real emissions are not published until later in 2015. Electricity producers' real emissions can vary if they purchase different types of renewable or CO_2-neutral power. Thus, the real total of DR's electricity use is not known at the time of the publication of the financial statements. The consequence is that Emission Details for 2015 are based on variable market data which can only be estimates from previous years instead of the actual year of publication. As a consequence, the stated CO_2 contribution of these financial statements can differ from the actual CO_2 contribution for 2015. This is the general approach for companies that publish this information at the beginning of the year, yet the use of these kinds of accounting mechanisms could be seen as a way to anticipate potential problems with year-on-year data not meeting the climate goals of the government.

The reporting of carbon data by DR is clearly more progressive than NRK and, similarly, the emphasis on infrastructure is justified according to estimates that measure electricity consumption at 51.5 per cent, heating of buildings at 12 per cent, and disposal and treatment of waste at 2.2 per cent of the total emissions of the organization. Other types of activities, mostly to do with scope 3 emissions outside of the direct control of DR, account for approximately 15.7 per cent. Significantly for us, transport and fuel consumption and process-related emissions on productions constitute 5.5 per cent (ibid.: 25). This is a relatively small figure compared to most of the other infrastructural areas of the operations required to run DR. However, it does need to be remembered that these production-related areas are, ultimately, inherently connected to strategic and material concerns for the company. Accordingly, focusing on the emissions from production is not only an operational concern as changes to regulatory or economic requirements will impact on material and strategic areas – that is, it influences the core competencies of the company, especially if these reporting requirements take on a more mandatory tone. And this argument does not take into account that commuting by taxi or car and transport by plane and train (key factors for performing the core reporting competencies of PBS) constitute a total of 12.5 per cent of DR's operations – coupled with the process-related emissions, these total 18 per cent of all emissions (ibid.: 25). If environmental considerations are now of considerable significance to the infrastructure and the performance of tasks related to core competencies of these companies, by this argumentation, an organization's environmental performance should be, at the very least, to be considered strategic – that is, as having potential major ramifications for its legal or other operational areas.

However, this is not something that has been applied fully to internal policy as DR's environmental guidelines feature little media-specific content. Furthermore, when material conditions of production are discussed, such as replacing older technology with ones that use less energy or conventional light sources with LEDs, they are only a small part of a much more wider focus on financial or societal sustainability. In this document, environmental issues total only two pages out of eight devoted to sustainability. One potential reason for these omissions is the coordination of environmental sustainability within the organization. Both NRK and DR rely on a committee that designates and oversees environmental responsibilities and operations. While such organizational oversight is important in consolidating policy, the example of BAFTA reminds us that a senior or a line producer needs to be involved in establishing and carrying them out at the level of production practice. The NRK and DR Secretariats, perhaps understandably, work at the level of organizational policy and view environmental sustainability as precisely that – as part of the management of the operations of a large organization. Consequently, there are few clearly specific protocols for media production in place. While most of the emissions are generated by infrastructural concerns, it would also be necessary to manage the core competencies of these companies in a much more direct way, as has been done by the BBC, for example. These issues will become more obvious when we explore an organization that has, from comparatively early on, adopted these practices: Yleisradio (YLE, the Finnish Broadcasting Corporation).

YLE: navigating the organizational management deficit

Environmental management at Yleisradio (YLE) follows patterns identified at DR and NRK, though the company has focused on developing much more in-depth guidelines than its Nordic compatriots. While a considerable part of this activity is concerned with creating awareness of environmental issues through content, YLE has also focused on environmental CSR since the early 2010s. They argue that this is part of being a public organization as 'responsibility is at the core of public service' (YLE 2014a: 11). For example, YLE published an Environmental Responsibility Programme document in 2014 (YLE 2014b) focusing on evaluating their environmental role. This is a comprehensive study of many of the different ways in which the organization's operations use resources. In comparison to NRK (which only has a few short paragraphs on the topic) and DR (with eight pages on general concerns and two on the environment), the length of the YLE document (21 pages) is already indicative of the significance afforded to the topic. The YLE strategy meets all the requirements we have seen with NRK and DR as its strategy aims to advance environmental activities through more efficient use of energy and recycling, transportation, procurement, waste management and education of staff to become more environmentally aware and capable. The total of YLE emissions in 2015 was 12,204 CO_2 tonnes, but in total the use of electricity has reduced from 34.6 million kWh in 2007 to 27.3 million kWh in 2015. Some 90.7 per cent of the waste generated is used for recycling and only mixed and dangerous waste is not processed through their environmental protocols (YLE 2015).

These figures are impressive, but they do not ultimately tell us much about how YLE operates as a media company. To counteract this suggestion, YLE indicates that the focus on environmental sustainability is essential as 'responsibility concerns all of the choices and decisions YLE makes as a media organisation' (ibid.). In contrast to NRK and DR, environmental responsibility is a substantial focus as their guidelines commit upper management, supervisors and staff to achieve environmental goals. The actual conduct of these operations is done through internal work clusters who contribute to strategic development. Here, the environment steering committee supports environmental responsibility and is comprised of supervisors from different parts of the company, including both areas of core competence as well as those with a more infrastructural focus, such as office management and building protocols. Organizational management is made more effective by identifying and naming the individuals in charge of particular areas of environmental strategy. These range from infrastructural specialists, such as the heads of services and of procurements to heads of development for productions and for sustainability.

The allocation of responsibility to specific departments and individuals means that YLE avoids the infrastructural deficit of many other media organizations. Much of the argumentation outlines particular production roles and technical issues, which must be incorporated with hands-on management ['Supervisors must have the tools to lead, maintain and oversee environmental practice' (YLE 2014b: 21)]. These strategies work to normalize the ecological habitus of YLE who 'will facilitate the best possible conditions for staff to actualise environmentally friendly practice' (ibid.: 21). Establishing environmental practice as daily routine contributes to YLE's organizational goal to act as an industry leader in environmental sustainability. Routine administrative tasks can transition to environmental media production practice in areas including travel (they discuss low emissions vehicles as well as its 32 environmental hybrid cars in addition their '20 turquoise YLE cycles for work drives' (YLE 2015)). Other directives include powering off computers resulting in 'savings of tens of thousands of euros a year' (YLE 2014b: 8). Ultimately, these are all intended to establish environmental sustainability as an everyday practice: 'YLE encourages its staff to make environmental choices in everyday work and expects its collaborators to agree with these principles' (ibid.: 3).

This emphasis on focused management drawing on the allocation of key roles continues with policy concerning the specificity of the emissions, or of how they relate to the production of media content. One of the company's top five priorities is the development of environmental guidance for the audiovisual sector (with energy consumption, car journeys, air travel, mixed waste comprising the other). This suggests that the company goes beyond the usual infrastructural protocol to develop enhanced media strategies. For example, one section of the document focuses on production lighting and how innovative technologies can be used to decrease the footprint (ibid.: 8–9). They explain that LEDs are not the most conducive practice for these purposes as their implementation would require early retirement of the existing lighting equipment. Another complication here is that the procurement of LEDs can be expensive and the amount they are able to project in illuminating sets can be insufficient if there is not enough investment in the volume of equipment.

While technology has improved and become less expensive since 2015, the framing of this debate is telling as YLE's (2015) report indicates that the percentage of LEDs will be increased where possible (YLE 2015). In comparison to some of the sustainability protocols of BAFTA, for example, YLE is lagging behind in applying technological developments to all its production practices. Similarly, they suggest a limited understanding of production arrangements with, for example, reusable cups on sets. These are in daily use but for large gatherings, studio or outside productions, these are not considered 'pragmatic' (YLE 2014a: 19). While this shows a level of development that does not meet the standards of BAFTA, it still emphasizes that media require its own set of material considerations where production strategies are evaluated on the contextual requirements of media production, not only on large-scale infrastructural areas.

In contrast to the other Nordic broadcasters, YLE provides a clear correlation between operational uses of resources – such as to do with buildings or water – and reductions that can be made directly from production practice. While locations such as the Metropolis at Tampere and the HQ at Pasila are decreasing consumption, YLE aims to cut back on its operating locations by 40 per cent by 2020. This reduction has two impacts on media production. The first is to do with distance working, which is considered an infrastructural practice, especially when focusing on administration. Simultaneously, this is also relevant for cutting emissions from post-production as specialists and journalists can complete work outside of the office, cutting back on transportation but also on operational costs. Another concern is the use of lighter equipment on sets as these not only cut back on transport emissions but also allow for smaller or local crews to conduct production practice. Such reductions are important but, at the same time, a significant part of the core competencies of broadcasting relies on producers, reporters and crew travelling to locations. These include the reporters' research done at external locations and journeys to international media events (such as for sport or political news). YLE calculates that 15 per cent of the total emissions is spent on work travel, with flights amounting to over half of this total, with personal cars at a third. The calculations do not include journeys to and from work (which would be more infrastructural in any case). A policy on travel is thus required to curtail the inevitable emissions that media companies have based on their core competencies.

Organizational management of external networks operates from a similar balancing between infrastructure and media specificity. Most of the procurements of vehicles and accommodation are sourced through Hansel, the procurement unit for the Finnish government. *Procurements Guidance* from 2012 shows how YLE's internal policy follows the State's Principality Decision on procurements, and also outlines how media procurements operate on the basis of particular requirements for media production (YLE 2012). An example used by YLE concerns the purchase of textiles and set materials to meet the requirements of production crews. These are commissioned through competitive mechanisms that are evaluated on the basis of environmental qualifiers including green certificates and lifecycle assessments of the materials and are, furthermore, based on the guidelines from the Global Reporting Initiative (GRI) as well as its GRI Media Sector Supplements.

The GRI (2014) is a significant addition to overcome the infrastructural deficit as the Supplements sector contains in-depth guidance bespoke for the media sector. The requirements necessitate in-depth reporting on all areas of production including specific measures on carbon intensity of papers, total volume of inks, glues by type, plastics used in packaging, specific batteries, and so forth. While much of the in-depth guidance is derived from infrastructural concerns, a commitment to the GRI venture requires reporting on these areas from the perspective of media production. In practice, this means the evaluation of production requirements for specific needs (for broadcasting, this would predominantly focus on studio conditions such as procurement of electricity and equipment or the use of LEDs or materials for set construction). Thus, the emissions of this sector are identified with a tangible footprint, from paper to servers, from lighting to batteries, and thus not only connect with an external infrastructural network but one that focuses on the specific interests of YLE.

Another area where this distinction becomes acute is in digital production and distribution. YLE makes the point that the supply and value chain of electronic and digital media has not been fully studied. Thus, one of the strategies developed by them is to explore how infrastructural and production emissions can be controlled with the virtualization of servers which could translate to a decrease in on-site electricity use and an increase in operational capacity. Yet, as we have suggested above, the use of cloud services comes with their own contributions to overall emissions. Reporting of these emissions is exceedingly difficult as measuring the impact of transmission and distribution of content involves a range of stakeholders outside of the company, for whom few applicable tracing mechanisms have been developed. At the same time, the diversification of the core competencies of a contemporary broadcasting company – which now include areas such as multichannel HD distribution and production of content for diverse platforms – complicates addressing the long-term impacts of production and increased consumption for a company like YLE. Both increase the need for storage space and capacity as well as the amount and type of devices receiving content, contributing to increasing energy consumption in the production and distribution chain as well as resulting in the increased proliferation of electronic waste.

Thus, when considering, for example, the Scope 3 emissions from the supply chain of its extended network, YLE would need to focus actively on communicating its environmental principles not only as infrastructural requirements addressing particular suppliers – of energy or server space, for example. While these would conventionally comprise infrastructural emissions outside of the direct remit of the company, with digital, they become part of the core competency framework. They would be strategic priorities for a company like YLE and impact on their ability to perform their operations and meet their financial imperatives. Strategies for encountering them would thus have to be organized from a perspective that takes into account the KPIs of what is effectively a multimedia organization. If prices for renewables can outpace traditional energy sources, or if purchasing services from Google using more efficient cooling methods in their Finnish centre, for example, is more cost-effective for managing YLE's digital operations, then environmental priorities would become strategic concerns for corporate management, and thus, an area that would impact on key aspects of the corporation's performance. For now, environmental sustainability is starting to emerge as a priority, but the strategies in place would need much more attention on integrating infrastructural advances with the core competencies of YLE's operations.

Despite these often comprehensive policy developments, the operations of YLE are not much different on the whole from the other Nordic broadcasting companies, and include a range of sensible, if expected, solutions for evaluating their footprint. Yet, the scope and implementation of the YLE strategies are a lot more advanced than those of DR or NRK.

YLE is also in the process of testing environmental protocols for the industry with the idea of developing a guide based on the BBC's production notes on sustainable productions. Most importantly, they show that YLE has approached these themes from a more focused sector-specific production perspective that ensures that the policies not only work as a CSR tool or as a PR exercise for public broadcasters to meet the environmental mandates and parameters set by general governmental policy. A striking area of YLE's work on this front is the level of transparency it maintains as this ensures that material considerations are appropriately considered and the policies are developed thoroughly with an appropriate level of input from both senior management and specialist members of the organization. Yet, if these goals are not addressed fully, sustainability documents outlining brief general policies may backfire on the companies in question – as we will now see with the Swedish broadcaster, Sveriges Television.

Sveriges Television (SVT)

The next section will switch focus from exploring corporate statements to an internal dialogue gone public regarding the CSR programme of the Swedish public broadcaster, Sveriges Television (SVT). This discussion will explore the balance between public service mandates and the realities of working in the media industry where sustainability initiatives are consistently perceived as potential obstacles to efficiency and costs. The focus on practical management of these strategies provides a necessary counterpart to the above exploration of broadcasting policy. The discussion was initiated by an article by Calle Elfström, an environmental journalist working for the organization, entitled *SVT on Its Way to a Climate Flop?* (Elfström 2016). In it, he critiqued the company for its promotion of a green image of itself while ignoring most of its internal regulations on sustainability. At stake here is the accountability of a public service broadcaster addressing the difficulty of meeting its strategic and financial bottom lines while introducing new environmental policy.

Before exploring all this, it is worth briefly addressing what the SVT environmental policy contains and how it positions the company's aims in both ideological and material terms. As with the other Nordic broadcasters, SVT is run independently from the state and from commercial interests but has to abide by general regulations for public companies as well as for large-scale operations. The SVT Environmental Policy (SVT 2009) replicates government strategies including *The Environmental Quality Objectives* that set protocols for areas like transportation and renewables. Most of the policy documents discuss key emissions areas to ensure responsible procurements across the company. Long-term plans include minimizing the environmental impact of their own programming and the carbon trace of their external collaborations on commissioned productions and services. The company has implemented a range of measures to conduct wholesale analysis of the carbon footprint of production processes and the purchasing of goods and services. SVT's rhetoric very consciously frames its operations as a climate-conscious organization: 'environmental work is integrated throughout the business and we continuously raise the ambition of our work' (ibid.).

These arguments are not much different from the general outlines of the other Nordic broadcasters. SVT balances between sectoral and infra-structural rhetoric in brief statements such as 'it is hard not to acknowledge certain conditions that exist when producing television. Often travel and energy-intensive technology are natural components of productions' (ibid.). Yet, some areas touch on a potential strategy that has not been considered by other broadcasters, especially the role of offsetting to compensate for large emissions such as energy purchases or air travel. Offsetting is a particular reterritorializing process as it disassociates the footprint of a production from its inherent materiality by substituting material costs for intangible economic trade-offs. Such practices are often favoured by organizations aiming to meet carbon targets but they have also met with considerable criticism especially in and against the affluent Nordic countries claiming global sustainability leadership. The maintenance of Norway's high GDP, for example, relies on substantial income from oil and negation of the environmental costs of fossil fuels through offsetting practices. Sweden is much more proactive in reducing emissions via tangible processes. As the OECD suggests, it relies much more on addressing the root causes of emissions through taxation regimes, for example, and is 'one of the few countries to successfully undertake a green tax shift, reallocating the tax burden from labour to environmentally harmful activities, namely CO_2 emissions' (OECD 2014). As this is a key government policy and as there is an emphasis on taxing the emitter, SVT must take heed of its public mandate. Thus, it is not surprising that SVT has made a special case of not adopting offsetting as part of its 'principles' (SVT 2012). The emphasis on dismissing this as a viable strategy is indicative of how public accountability of these organizations limits the use of translational tools used to displace materiality into reterritorialized statistics.

But this can also be considered from an alternative angle. The emissions from broadcasting are considerable and would require substantial funds if offsetting was considered a viable strategy. To take an example, YLE's 12,204 tCo_2 for 2014 would translate into offsetting costs of £73,230, which would influence its budget considerably. Offsets are often done in substantial volumes by large commercial corporations when they are forced to meet climate targets. Yet, broadcasting works under the general climate goals of the state and is predominantly funded by taxes. The use of tax money on offsets could be seen to contradict the necessity of a broadcaster to focus on cutting back its own emissions through internal policy, especially as these emissions are often not considered to be part of the core competencies of these organizations. And even if they eventually become strategic priorities, as with the emissions from digital infrastructure, operational policy on procurements would likely be the first method to influence these emissions.

Abiding by its regulatory mandate ensures the importance of sustainability protocols for production policy, but as Jimmy Ahlstrand, Strategy Director for SVT, suggests, the inclusion of environmental sustainability in company policy

> is primarily for economic reasons. Instead we have in this context prioritized reduction of resource consumption (instead of focusing on buying carbon offsets, our discussions tend to focus on how to, for example, minimize transport or focusing on train journeys instead of air).
>
> (Ahlstrand 2016)

Reduction of consumption makes a lot of sense as it prioritizes actual reductions instead of the reterritorializing tactics of offsetting, but as it seems sustainability is integrated into policy for economic reasons (especially as transportation is included as a key cost – cutting car journeys, for example, will inevitably cut costs), SVT's strategies need further exploration concerning the company's climate goals.

Calle Elfström's article (Elfström 2016) published in the company's internal magazine *Vi på tv* provides this material. The piece suggests that SVT's goal of halving its emissions from 2007 to 2020 has not been met. One of the key concerns is that most of the staff working at the company do not even know it has a set of climate goals. While the role of SVT in communicating environmental content is widely promoted, its attempts to curtail its emissions are not widely discussed or acknowledged. These omissions in communications – suggesting that they are not a priority – are complemented by Elfström's suggestion that most of the existing rhetoric is simply words for promotional purposes: 'if you count optimistically, SVT has been able to cut emissions by 6.5% since 2007 but figures and patterns go up and down' (ibid.). This lack of consistency is identified as a problem as various types of business journeys make up a large part of its total emissions (83 per cent, which the company calculates as part of its direct emissions, meaning they do not take into account personal travel to and from work). Air journeys for 2014 total 232,000 kilometres which is 580 times around the world. Long distance travel makes up 85 per cent of these journeys. One of the ways that the company aims to cut this is by replacing air with train on short journeys of less than 50 miles. The problem, for Elfström, is that there is no concrete advice for staff on how to plan for and organize their travel to take into account efficiency and cost as well as a lack of tangible guidance on which types of air travel to cut. Elfström argues that most of these journeys are entirely necessary for news programmes and other production considerations, but as there is no clear guidance or transparency on how to achieve them, the statements come across as empty rhetoric designed predominantly to tick boxes.

The brevity of the policy comprises another problem as it only has seven lines on climate goals and no conclusions on whether the company will meet or miss its aims. SVT has commissioned a separate climate report from Tricorona, a consulting company, but this, according to Elfström, does not provide a full picture. Tricorona claims that reductions from the previous year 'must be considered very good' (ibid.), yet they suggest that SVT has not set a base year for analysis. Elfström argues that this is clearly signalled as 2007 in the climate report which Tricorona 'must not have read'. He argues that emissions have been increasing year on year and this rhetorical obfuscation would be called greenwashing in SVT's news reporting on other companies. The suggestion is that if someone is paid by a company to do a report on their environmental role, they will not be 'particularly critical' (ibid.). Considering Elfström's role as an environmental journalist, it is understandable that he would evaluate the operations of the company for whom he works from such a critical vantage point, especially as the transparency policies of public service broadcasters emphasize that they must be forthcoming on how they meet societal goals. Elfström's key argument is that if it is SVT's role to scrutinize other companies for their environmental role, it is essential that they must uphold the highest of standards in their own organizational management.

The article led to a reply from Strategic Director Ahlstrand entitled 'We have begun reviewing our current objectives' (Ahlstrand 2016). This was also published in *Vi på tv*, and aimed predominantly at internal readers. Pointing out that SVT has issued an environmental report since 2007, Ahlstrand outlines the ways SVT has met standards in energy electricity and fuel use, reducing their environmental impact by 20 per cent in 2010 alone. A shift to an energy mix highlighting renewables is credited as the reason for this achievement, indicating that SVT is meeting its environmental quality objectives. Other practical means have been implemented, including more efficient uses of space and an increase in the use of environmental cars. However, these strategies do not tell us much about how SVT is addressing its core competencies. As we have argued above, they are an essential element for media companies to address when identifying how their operations reduce emissions in the long run. Ahlstrand also notes that the company switched its environmental accountants in 2011, arguing that Tricolora, the consulting firm critiqued by Elfström, has much more effective measurements than the previous operator, but this comes with the caveat that they can only provide an emissions trajectory from 2011 onwards. With the new measurements, SVT has been able to minimize its emissions by 9 per cent in total between 2011 and 2015. Ahlstrand notes that this may appear lower than in the previous measurement period (20 per cent reduction in 2007–2010), but the variation is to do with the different measurement systems as Tricorona's measurement protocols are more accurate and thorough and inevitably lead to a decrease in the reductions percentages.

While this argument may make sense from a management perspective, it indicates a similar rhetorical approach as was the case with DR raising questions over reporting standards. These uses of reporting mechanisms as discursive devices to suggest the possibility of flexible interpretation provide a valuable reminder that the indicators for environmental veracity, cost effectiveness, and so on are not only management tools. They are also indicators of the anthropocentric role of sustainable media management as they both reterritorialize environmental emissions and organize operational networks in ways that prioritize organizational efficiency over environmental sustainability. Of course, it would be completely naïve to assume any other priorities for media businesses, but as Elfström suggests, there are important questions at stake here about integrity and the role of CSR in the organization. Ultimately, these rankings are based on comparative statistics with similar organizations where the purpose of ranking is attuned to demonstrating value systems while highlighting operational efficiency.

As Ahlstrand has suggested, environment management is useful in contributing to the economic and operational efficiency of SVT but has considerably less use value as a material part of the company's operations – it could continue to conduct its operations as before even if it failed on its environmental policies for its production emissions as general mandates would be met through infrastructural transformations. For Ahlstrand, this is about adjusting expectations on what a sustainability policy can achieve: 'perhaps the company has set its goals too high, but this is better than a lower goal or having no goal at all' (ibid.). The position of environmental sustainability as part of the organizational priorities of SVT is thus relatively marginal, even though, according to Ahlstrand, SVT has received new guidelines from the board in June requiring a clearer environmental policy. This has led to a review of all aspects of policy, and information on climate actions is 'freely available in the climate and annual reports'.

The discussion was picked up by the CSR specialist magazine *Aktuell Hallbarhet*, who interviewed Elfstrom about his article. In reply to Ahlstrand's piece, Elfström questions transparency when 'information beneficial to the company lies in someone's drawer' (Elfström, cited in Rosengren 2016a). While the Climate Report and the environmental policy are available, versions published on the intranet 'lack information and clear conclusions'. As Elfström's article criticized Tricorona for being 'too kind' in its work on assessing SVT's performance, the CEO of Tricorona, Christian Patay, wrote a response suggesting that Elfström misunderstood the implications of SVT's requirements from the company (Patay, cited in Rosengren 2016b). According to him, Tricorona's task was to provide a calculation of SVT's climate impact and not supply tips and advice for the company on how to improve its performance. Consequently, accusations of greenwashing are invalid as the positive tone of the report is a result of its basis as a calculation, not an evaluation. Patay argues that the positive judgement of SVT's operations was also included under the subheading Flight Trips and covers only flights. As this was an area in which SVT has been relatively successful, it was taken as a case study.

After the reply was published, Elfström issued a corrective to say that 'no shadow may fall' on Tricorona (a phrase replete with a sense of sarcasm) (Elfström, cited in Rosengren 2016a) but the problem remains concerning SVT's ability to provide concrete data on where it has improved and where it will improve its operations. In any case, Elfström agrees that a review of sustainability policy may work to increase SVT's sustainability measures but transparency and accuracy remain concerns (ibid.). For example, Ahlstrand suggests that SVT 'compares favorably with other service providers' as environmental consultants call the emissions 'reasonable relative to an office-based operation' (Ahlstrand 2016). Elfström's view is that this is not enough as a public service provider must be better than other, especially commercial, organizations (Elfström, cited in Rosengren 2016a). Furthermore, the lack of sector-specific goals is an issue that undermines this perspective on the validity of the existing measures. As these goals tend to be difficult to develop or they are not seen as significant enough for financial or managerial requirements, they are often marginalized, as seen with other public service broadcasters like DR and NRK. More concrete strategies are required, especially for companies priding themselves on their generation of environmental awareness and their CSR operations. Infrastructural regulations often compensate for the lack of any sector-focused policy (or are used as such), which makes sense when we are looking at cultural policy (of which CSR is often considered a part). Cultural ministries set the guidelines for domestic policy on areas such as journalistic ethics and diversity which are taken up easily as they are core competencies for the media. Environmental content also falls comfortably here. Yet, the idea of environmental management is a much more dissonant proposal as it is an operational concern (to use the Media CSR Forum terminology) and accordingly more of a PR exercise. Thus, Elfström suggests that perhaps there was interest in the topic some years ago, but since then the interest has decreased (ibid.) due to certain changes in staffing as well as in the priorities of management. As the role of sustainability management does not meet the company's perception of its core competencies, it is often overlooked as a KPI. Accordingly, in a reply to my requests to access to these reports, Ahlstrand suggests that SVT is 'currently reviewing our environmental policy and this might lead to changes in our reporting. Currently we are doing a yearly measurement of our environmental footprint based on CO_2 footprint' (Ahlstrand 2016).

Conclusion

Environmental policy for broadcasting is complicated by two issues: infrastructure and core competencies. In terms of infrastructure, it is worth remembering that over 80 per cent of energy consumed by these companies tends to come from electricity, heating and distance cooling.

Production-related emissions are comparatively small, but still significant. Yet, the role of infrastructural requirements overshadows these, arguably, due to the organizations' close link to government mandates and funding. Infrastructural requirements also tend to correlate with intergovernmental regulations, indicating the lack of thorough regulation domestically or at the international level. For example, the DR strategies outline how its catering adheres to EU directives on the necessity to aspire to being 30–50 per cent ecological. Facility management such as cleaning services are based on EU directives on waste and restrictions on recycling of paper, biomass, etc. These links shape the approach of these companies to CSR but also show that replicating frameworks can be productive but also somewhat limited in their scope and productivity.

The responses of the companies to their footprints veer from replication of infrastructural standards to specific tactics for curbing emissions. NRK, DR, YLE and SVT show how environmental decisions can rely on state infrastructure to different ends with some avoiding these issues through the responsibility deficit, while others have started to work on the difficult task of evaluating emissions relating directly to their core competencies. But they also show that the companies with the most advanced strategies tend to be proactive in their own right and take into account the fact that the broadcasting sector has a different set of requirements not only from other media but also from the more generalized guidance for office-based sustainability management. Replicative networks tend to be weakest in terms of understanding the impact the sector has, as they predominantly rely on infrastructural concerns without paying attention to the core competencies of the sector. The strongest networks come from conditions where connections are not built on pressure to meet domestic climate goals nor from general societal legislative directives. The cases of YLE, and to some degree SVT, show that a more emphatic role on engaging stakeholders and potential participants internally and externally can shape company policy and lead to both sustainable practice and increased efficiency.

Bibliography

Ahlstrand, Jimmy. 2016. 'Vi har inlet en overseen av vara nuvarande mal', *Vi på TV*. Available at: http://vipatv.svt.se/204/debatt/arkiv-for-debatt/2016-09-28-vi-har-inlet t-en-oversyn-av-vara-nuvarande-mal.html (accessed 11 November 2017).

DR. 2014. *Årsberetning*. Copenhagen: DR.

DR. 2015. *Årsrapport*, Copenhagen: DR.

Elfström, Claes. 2016. 'SVT på väg mot klimatflopp?', *Vi på TV*. Available at: http://vipatv.svt.se/204/debatt/arkiv-for-debatt/2016-09-28-svt-pa-vag-mot-klimatflopp.html (accessed 11 November 2017).

GRI. 2014. *G4 Sector Disclosures: Media*. Amsterdam: The Global Reporting Initiative.

NRK. 2014. *Årsrapport*. Oslo: NRK.

NRK. 2015a. *Langtidsstrategi*. Oslo: NRK.

NRK. 2015b. *Årsberetning*. Oslo: NRK.

NRK. 2016. 'NRKs arbeid med samfunnsansvar'. Oslo: NRK. Available at: www.nrk. no/informasjon/xl/nrks-arbeid-med-samfunnsansvar-1.13193070   (accessed   11 November 2017).

OECD. 2014. *Environmental Performance Reviews: Sweden*. Paris: OECD.

Rosengren, Lina. 2016a. 'SVT anklagas för greenwashing av sin egen reporter', *Aktuell hållbarhet*, 30 September 2016.

Rosengren, Lina. 2016b. 'Tricorona svarar på SVT-reportens kritik', *Aktuell hållbarhet*, 3 October 2016.

SVT. 2009. 'Företagskultur'. Available at: www.svtb2b.se/?page_id=72 (accessed 11 November 2017).

SVT. 2012. *Att producer för SVT: en guide för bra samarbete*. Stockholm: SVT.

YLE. 2012. *Vastuuraportti*. Helsinki: YLE.

YLE. 2014a. *Yleisradion hallintoneuvoston kertomus eduskunnalle yhtiön toiminnasta vuonna 2014*. Helsinki: YLE.

YLE. 2014b. *Ympäristövastuuohjelma*. Helsinki: YLE.

YLE. 2015. 'Huolehdimme ympäristöstä'. Available at: https://yle.fi/aihe/artikkeli/ 2015/10/15/huolehdimme-ymparistosta (accessed 11 November 2017).

YLE. 2017. *YLEn vastuullisuus*. Helsinki: YLE.

8 The publishing industry in the Nordic countries

Introduction

Environmental management of different media – here, broadcasting, publishing and film – is a complex practice that requires an understanding of regulatory incentives as well as industry practice. As we have seen, this tends to be at its most effective when managers address and shape organizational policy based on the particular requirements of their sector. For broadcasting, this has been premised on fulfilling its public mandate and continuing to meet carbon targets set by the state on environmental best performance. Simultaneously, some of the more advanced policies highlight recommendations to evaluate the emissions generated by the core competencies of these organizations, including journalistic practices and production techniques. The circumstances and particularities are different for private companies who still abide by most of the general infrastructural legislation but are not bound by, for example, principality decisions on procurement as the public broadcasting service (PBS) tends to be. Thus, any attempt to rely on infrastructural arguments must be interrogated closely in relation to the modus operandi of each company in its specific media sector and its wider political economic context.

If broadcasting struggles to develop environmental sustainability management protocols, largely due to both the responsibility and infrastructural deficits, private commercial publishing companies have been early adopters of environmental practice. This is not surprising considering the environmental impact of the core competencies of the sector. While broadcasting is often perceived as 'invisible', at least for general audiences, print leaves the types of tangible material traces that can be observed in both daily life and environmental politics. For consumers, the continuous flow of messages concerning the necessity of recycling paper and the amassed piles of newspapers and magazines visible in daily life serve as constant reminders of the materiality of print media consumption. For regulators, the publishing industry, especially print, is an area that can be visibly and effectively regulated due to environmental priorities including policies on foresting and curtailing the toxification of waterways, both especially prominent concerns in the Nordic countries. Publishing also provides room for innovation as the role of digitalization is pervasive in the publishing industry.

This chapter discusses the environmental sustainability management structures of two large Nordic presses, Alma Media and Schibsted. Both publishers have implemented extensive environmental accounting measures into their daily operations – both at the level of infrastructure and in core competencies – and have thus received commendations for their work from international sources such as *The Guardian*. They also play a founding role in the Media SCR Forum, contributing to establishing strategies for environmental management of diverse types of media. The chapter will investigate the rationale of these companies in adopting environmental accounting and management, exploring to what extent this adoption is a reputational and economic imperative designed to cut back on costs and increase their brand image, or whether this is to do with establishing the companies as leading innovators in policy and technology.

Regulatory framework

The obvious starting point for exploring the publishing industry comes from the infrastructural procurement guidelines set by the various governments. Print and paper consumption are included in ventures like the WWF's Green Office which provides best practice guidelines for most office operations from procurement of paper and pulp sources to recycling. The Nordic countries, especially, have been at the forefront of many of these developments due to the reliance of their economies on the foresting industry, especially in the case of Finland and Sweden. While this does not translate directly to regulatory oversight of the core competencies of publishing companies, it forms a key part of the environmental protocols established by the respective Ministries for the Environment, and thus influences the infrastructure of the companies' procurements but not necessarily how they meet their core competency key performance indicators (KPIs). To illustrate some of these distinctions in practice, pulp procurement would not impact significantly on content production at most publishing organizations, though other material requirements, for example, in the form of prioritizing digital platforms, would certainly do so.

Other regulatory frameworks would also influence environmental management at companies like Alma and Schibsted. Both are large commercial companies listed on the stock markets. Thus, they have to abide with the EU Directive 2014/95 concerning companies that have more than 500 employees on their balance sheets as these companies must, at a minimum, report on environmental and social matters in their directors' reports. In comparison to the UK Companies Act, the Swedish Companies Act of 2005 does not impose requirements on the same level as the UK in terms of environmental reporting. This is arguably due to an emphasis on consensus within the Scandinavian countries, where explicit regulatory regimes are often considered antithetical to normative social order. Accordingly, the Swedish Companies Act lays responsible 'duties on the board of directors of a company, including a requirement to act in the interest of the company to promote the success of the company to the benefit of its shareholder' (SFS, Swedish Companies Act 2005: 551). This is in line with some of the basic philosophies of the Scandinavian managerial system, as described by Eric Rhenman and Bengt Stymne (1965). They argue that the structures of organizational management in this context differ from typical neoliberal corporations in their societal focus. Initiatives start by identifying and responding to external needs instead of focusing squarely on corporate bottom lines and shareholder priorities.

While this organizational management structure, starting out from external needs and using them to meet internal KPIs, may sound somewhat utopian, it operates as a key principle for the 'Scandinavian Cooperative Advantage', defined as 'the general tendency for companies in a Scandinavian context to implement a value creating strategy based on cooperating with their stakeholders that results in superior value creation for the companies and their stakeholders' (Strand and Freeman, 2015). Many of the large media companies and their boards will consequently consider the impacts of their operations on the environment, society and social circumstances. Creation of such shared value is implicit in how these publishing companies position their mandates. Schibsted describes its mission as follows: 'We believe that the secret to our continued success lies in understanding, truly understanding, the future needs of our customers – all the millions of people that use our various services every day' (Schibsted 2014). Alma Media suggests its socially conscious work will enhance the lives of all members of society as it aims 'to promote democracy, free speech, reliable dissemination of information and the wellbeing of its reference groups' (Alma Media 2017a).

To achieve these goals from an environmental perspective, both companies start from infrastructural sustainability and organizational management. Buildings, sustenance, heating, travel and environmental practices among staff are all mentioned but as most large companies subscribe to these practices, it is much more productive to focus on their core competencies. Here, print – the visibly material component of publishing – continues to be a significant part of the business model, but they are increasingly turning to digital as their primary mode of production. Evaluating scope emissions is less problematic for print as measuring energy that goes into printing processes or the volume of, say, pulp and ink in newspapers is verifiable. When it comes to digital, tracing patterns of users, including how much time they spend on pages, is a much more difficult task. As we have outlined above, this does not account for the resources that go into the server and cable infrastructure enabling these communications, nor of the patterns of usership as well as the particularities of terminal devices. Calculating such a wide range of Scope 3 emissions, an area that is still largely out of the hands of publishing companies, is difficult, especially considering that it is in their interests to increase both the proliferation of content and the time spent on each page or ad. Yet, there is a general consensus that print and digital media currently exist in balance (admittedly one that is shaky and shifting). Print remains a key means of revenue for the majority of book and magazine publishing houses, but even as this continues to be the case, documents, such as the *Future Report* by Schibsted (published every year to take into account new technological and industrial developments), construct narratives that do not leave room for ambiguity over where the industry is heading. Another key indicator of these shifts is the increasing role of multimedia content that blurs the constitution of the medium. Both companies frame their work as those of multiplatform organizations, and, thus, analysis of these publishing companies needs to approach this technological, and material, transition as a key area.

Alma Media

The first case study is of the Finnish company Alma Media, one of the leaders in developing environmental sustainability for the European media industries. Alma Media has come up with a complex corporate social responsibility strategy (Alma Media 2017b) that relies on a somewhat orthodox understanding of the implications of sustainable development. This is based on three principles which are: (1) efficient operations; (2) responsible supply chain; and (3) increasing environmental awareness. These are aimed at minimizing negative environmental impacts on content and production operations. While CSR strategies include the obligatory emphasis on press freedom and diversity, environmental arguments are present in most of the documentation of the company. The significance of environmental practices comes through in the volume of coverage afforded to it as their annual review for 2014 devotes 20 pages out of 47 pages to sustainability (Alma Media 2014). The focus here is on the wide-scale adoption of green processes to rethink the operations of the company. The document makes this clear by explaining that 'by improving its material and energy efficiency' (ibid.: 40) the company can achieve a considerable level of environmental sustainability.

The key environmental impacts of Alma's business are related to printing, distribution, properties and travel. Many of the strategies deployed by the company reflect those in the business community. Operations focus on material and energy efficiency through new technological innovations such as a heat recovery system that captures over 80 per cent of the exhaust air in buildings. Alma buildings have received LEED environmental certification with their Töölönlahti headquarters the recipient of LEED Gold certification in 2013 in recognition of its environmental advances. This recognition was due to construction methods and technology highlighting the circular economy in both the supply chain and in continuous office management, including a heavy emphasis on renewables and recycling. While these are all infra-structural concerns, the company has integrated emissions from its core competencies into reducing its direct environmental impacts. These tactics include printing operations that cut the use of solvent-based detergents, emphasizing the need for water recycling and evaluating the maculature percentage, which expresses the proportion of waste material to total material used in printing. Environmental considerations also feature in purchases of newsprint pulp, chemicals and printing plates as, for example, in 2014, 75 per cent of the newsprint contained recycled fibre. In 2015, this had gone up to 85 per cent. They estimate that a single copy of a newspaper amounts to the emissions of a 1 km journey by car. The fibre of the newspaper can be recycled 4–7 times and its CO_2 decreases by 20 per cent when recycled. Differences between newspapers are caused by the different newspaper formats – tabloid and broadsheet – and the difference in weight due to the number of pages as well as differences in the amount of ink used. These reductions are significant for the publishing sector and arguably set it apart as a positive case for implementing environmental practice as part of its operations at the level of its core competencies.

While the code of conduct for management and staff starts out from anthropocentric managerial rhetoric, it provides ample space for agentic material to influence the constitution of the Alma network. Key responsibilities are allocated based on internal hierarchies, each layer with responsibilities meeting their particular sustainability KPIs:

> The guidelines for the Sustainable Media perspective are provided by the Group's executive management and the Board of Directors as a part of the Group's strategy work. At the core of the CR operations is the corporate responsibility network, which spans all of the Group's business units. The operation of the network is coordinated by Alma Media's CR manager. The network also develops the responsibility of each business unit in line with common guidelines.
>
> (Alma Media 2017b)

The description suggests a similar strategy for assigning responsibility roles as was the case with YLE. While these can be perceived as mere rhetoric to show off the company's credentials, Alma has taken the staffing of key roles for environmental matters even further. Roles such as the Quality and Environmental Manager of Alma Manu, the printing and distribution unit, contribute to the establishment of environmental targets for the company which emerge from specific operations of the unit. These sorts of roles do not exist in many other media sectors as environmental care here falls under the principal operational aspects of the company and is regulated by law. These roles are significant as the 'polluter pays' principle would be enforced on a company whose key outputs are potentially environmentally hazardous. While many other sections implement environmental policy actions through the supervisory roles they hold in, for example, management of company property and recycling operations, an interesting variation on this emerges with distribution as most of Alma products are delivered on foot or bike. If cars are used, specific training on economical driving techniques is provided. Delivery personnel are also asked to undertake waste management training for the newspapers, magazines and other content they distribute. While these have a more generalized function, they are also based on the core competencies of Alma as a publishing house. Thus, their roles include strategies to deal with, for example, minimizing VOCs (volatile organic compounds) in the dampening solution in the process of printing or the waste management of toxins and other harmful substances.

Prioritizing such roles is in sharp contrast to the public service broadcasting sector where general CSR managers would oversee the environmental operations of these companies. The description of the internal management network makes it clear that there needs to be strong central encouragement from the senior management, which draws on the KPIs of the organization as a whole. In the case of Alma, these would be its printing operations (and nowadays increasingly its purchase of digital services). They are then taken up by individual units, so policy can be implemented into particular material realities that matter for specific operations within the wider organization. These units act as important nodes ensuring that sustainability policy is successfully adopted to practice. Such arrangements are largely absent among the broadcasters, for example, as outside of YLE, these sorts of role allocations of particular responsibilities tend to be overlooked due to the somewhat immature stage of development of environmental sustainability policies, manifesting, arguably, in the dominance of the responsibility deficit in this sector. For Alma, these infrastructural strategies continue to be important and, significantly, in terms of the organizational KPIs of a media company, they amount to more than operational concerns. As part of organizational management at Alma, environmental sustainability is a significant KPI within its general policies on corporate sustainability. This is in line with the core competencies of a publishing company reliant on environmentally detrimental modes of production for its key products. Thus, for publishing, it is not surprising that environmental sustainability emerges as a strategic concern that influences most parts of organizational management at the company.

The emphasis on complex managerial oversight is reflected in Alma's development of global networks. They were the first Finnish media company to take on board the Global Reporting Initiative and its additional Media Sector Supplement. This has led to the publication of an annual corporate responsibility report since 2013 and participation in the 2014 Carbon Disclosure Project (CDP), a global initiative on climate impacts. In addition, Alma has been a signatory of the UN Global Compact initiative since 2011, being one of the first media companies to do so. Consequently, it has consolidated networks both internationally and across the sector to integrate its own sustainability protocols with up-to-date international standards. This has led to it ranking as the top Nordic media company for sustainability in the CDP. Other organizations it has collaborated with include the VTT Technical Research Centre in Finland and the Swedish Centre for Sustainable Communications, both on a study on the footprint of the company, completed in 2013.

It is clear that Alma has developed a comprehensive environmental policy that is internally and externally expansive. Yet, the key principles on which this ethical approach relies need further evaluation. The purpose of this is not to demean the efforts of the organization but to position them in the wider field of rhetorical and practical approaches to sustainability adopted by the media industry. This is especially the case as the rhetoric they use is closely linked to the ways sustainable development balances economic and environmental ideas:

> Sustainable media refers to responsibility in our own operations and making responsibility a fixed element in our business. Responsible operation and profitability are not conflicting goals. On the contrary, they are preconditions for each other: economically sustainable media needs to also be socially and ecologically sustainable, and vice versa. Changes in media consumption and consumer behaviour create a need for renewal in the media sector.
>
> (Alma Media 2017b)

The narrative we see forming here is one that emphasizes both innovation and increased efficiency based on the Scandinavian outside-in principles for responsible companies. The case for profitability is especially intriguing as Alma directly confronts the critiques of green capitalism often levelled at environmental sustainability initiatives. For them, environmental sustainability needs strong organizational support and a thriving organization needs to be sustainable – both economically and ecologically – to be able to compete in the contemporary marketplace. Simultaneously, sustainability can function as a means for innovation and enhancing the profile of a media company, and as we saw with its external networking mechanisms, Alma prides itself for being at the forefront of innovation internationally. In fact, Alma considers the integration of sustainability into its brand as an essential strategy.

Image management

Reputational management is vital for contemporary media companies, especially in the Nordic context, but we have to also remember that the Responsible Media Forum considers environmental sustainability – a theme with considerable PR value – as part of their operational value, and not as something that would be strategic or material for these companies. Alma documents, on the other hand, indicate that environmental sustainability plays a much larger role than mere PR, resulting, arguably, from sectoral differences within the media industry. If the broadcasting sector uses sustainability to fulfil some of its responsibility mandates, while simultaneously relying on regulatory incentives to avoid full-scale adoption, publishing sees the implementation of these strategies as a means not only to fulfil mandates, but also to promote its innovative capabilities. To understand the significance of this, we must consider how stakeholder value is created in the Alma network.

Kennon, Howden and Hartley (2009) argue that stakeholder networks can provide a range of advantages for understanding how the communications and operations of an interest group operate. To take an example, the use of certificates like the Nordic Swan are not only stamps of recognition but vital areas of image management. More importantly, they act as key nodes within the sustainable media network as they communicate in succinct terms to a variety of stakeholders – including the board of the company, staff, competitors, consumers, governments – about both the values and actions of that organization. These are especially important in the Nordic context where shared value and social relevance meet not only social and governmental expectations for the sector, but also have the potential to become business concerns. Heikki Masalin suggests a somewhat similar perspective where the general focus within the creative industries has been on stakeholders as key means for value creation in the creative economy which 'focuses on the creation of stakeholder value – shared value – [which is] any mix of social, cultural and economic value ... which bring arts, culture, technology and business together' (2015: 37).

Certainly, sustainability works as a key tactic for creating value for a range of stakeholders, acting as a mechanism for promoting efficiency through cost incentives and anticipating regulatory controls while also emphasizing sustainability strategies and the cultural and social capital they create for an organization. Yet, its key role for shareholders emerges from its image management potential and the material benefits this can provide. Sustainability strategies can be used to rebrand organizations, a factor we now illustrate by exploring how Alma is rebranding itself from a publisher to a digital services provider. This is a necessary strategy for Alma as it has to operate in a marketplace where competition is rife among not only publishers but also emerging platforms in retaining audiences and revenue streams. In such a context, the ability to innovate provides capital that enables companies to stand out and meet the KPIs of diverse stakeholders. Indicating potential for ongoing return on investment in such a competitive market, brands are used to 'cultivate and use more abstract sets of values to provide leverage for line extensions into new areas' (Moor 2007: 37). If we consider sustainability strategies as brand management, they allow companies like YLE (and to a more problematic extent SVT) to consolidate an impression of their social value. Even more so for companies like Alma or Schibsted, this enables them not only to consolidate their roles as beneficial information providers but also to branch out into innovative areas like developing emerging technologies and practices. Innovations in integrating sustainability signify to relevant stakeholders that Alma is at the forefront of responding to key emergent developments.

Considered from this angle, the central role of environmental policy at Alma appears as a means to gain reputational capital, providing them with a competitive advantage over other similar publishers in Finland and the Nordic countries. Whether this has any monetary value in terms of consumers is open to question, but for other stakeholders, including investors, these brand embellishments are important. In the contemporary ecological habitus, environmental labels for media production, including ones like Albert, are not often aimed at consumers but rather the industry. Professionals may have the level of literacy necessary to recognize these labels in the corners of packaging or in the end credits of a television show and be willing to pay attention to particularly innovative production processes. Thus, when specialists in marketing talk about the ways brands use the emotional or affective associations of consumers to build brand equity, a type of 'value', which, as we have seen, can be included on company balance sheets and can contribute to stock market value, the affective associations of the Albert logo or the Nordic Swan connect with professionals who recognize the brand as the embodiment of a new form of informational capital. This translates into real capital as Alma's top Nordic ranking in the Carbon Disclosure Project means that 'about 820 institutional investors around the world utilise the CDP results in making investment decisions. Investors consider good results an indication of the companies' good management of climate issues' (Alma Media 2017b).

Brand identity and management are an essential part of Alma's external networking. For example, in 2014 the company took part in the Ratkaisun paikka event (Decision Point, the leading CR event in Finland), with Alma as the main media partner. The company used this platform to present its sustainability strategies to promote corporate responsibility thinking in Finland, especially on the role of ICT. Other forms of collaboration have focused on more local operations such as the City of Helsinki Climate Partners network. This is an event aimed at 'promoting cooperation to reduce climate emissions and boost the competitiveness of the participating companies' (Alma Media 2017c) who operate in and around Helsinki. Once again, the active role of the company in these networks increases its reputational capital, further enhanced by its involvement as the main partner in the Responsible Summer Job 2015 campaign for Helsinki, and development of Alma Media's Sustainable Media programme through co-operation with the Climate Partnership Network, a cooperative network for Helsinki and its businesses.

In addition to its ranking as the top Nordic media company by CDP, Alma has collaborated with Eurosif, an umbrella organization for socially responsible investment in Europe. Eurosif conducted a study of the media sector in 2012, in which Alma was one of the case studies of successful sustainability management. While presentation of the case study focuses on journalistic principles, the document also contains an interesting outline of business risks and opportunities whereby branding is identified as a key area through which environmental issues can improve the company's performance: 'A focus on environmental issues (in operations and programming) may provide companies with cost savings through: reducing the use of resources and energy, brand benefits by allowing them to market their credentials and better customer relationships' (Eurosif 2012: 2). Environmental sustainability is thus an essential part of the rebranded Alma which needs to 'efficiently decrease its ecological footprint by changing its operating methods' (Alma Media 2014: 41). The emphasis on the company's ability to adapt and innovate is significant here as they form part of the core competency framework Alma is trying to sell to its stakeholders. The conflation of sustainability and digital operate as two key parts of this branding exercise. The strategies start out by identifying the transition to digital as a particular area for environmental efficiency. This argument is largely premised on the notion of decreasing emissions through switching from paper to digital. For the company, 'the number of pages printed has decreased by 40% to 3.2 million pages per year'. Yet, simultaneously, 'while ICT equipment and infrastructure have significant environmental impacts, careful analysis and mitigation of these impacts are of great importance as digital media consumption increases' (ibid,: 41). Thus, connecting innovation in both environmental measures and digital technology can allow Alma to suggest that as they have been able to transform their core operations to incorporate environmental sustainability on all levels, they will also be capable of expanding from traditional publishing operations to a digital multi-platform operator, as well as all the challenges this brings for its current environmental strategies.

Digital emissions

One of the key arguments of the Eurosif study is the need to be aware of the substantial risk of the changing landscape of media production and consumption for both the company's profit margins as well as its environmental impact. The study warns that media companies do not consider terminal devices to be their responsibility and instead perceive them to fall under the mandate of the telecommunications industry. With the convergence of both technological platforms and media sectors, these may turn out to be future risks for Alma. Branding itself as an innovator, it makes sense for Alma to work with its partners to ensure that these risks are anticipated, especially as Alma as an organization is now enmeshed in technological convergence on most levels of its operations. Reflecting the challenges facing the publishing industry, Alma aims to rebrand itself as a service provider with four business units: Alma Markets, Alma News & Life, Alma Regions and Alma

Talent. Alma Markets (previously Digital Consumer Services) includes online services from housing to recruitment while the other three provide both print and online publishing services for a wide range of magazines and newspapers. For example, the 2014 document outlines how Alma Regions can continue to publish newspapers by subsidizing them through innovative digital content and services.

Digital operations are, thus, an integral part of the development of operations at Alma and connect concretely with the company's environmental goals. The problem is that there are no adequate measurements in place to evaluate the long tail of emissions from digital production and consumption. Alma suggests that the widely held assumption that digital media is more beneficial for the environment persists for both industry and consumers as well as in the attitudes of advertisers and NGOs. Having 'decided that it was time to find out what the reality is' (Alma Media 2015), Alma commissioned a study from VTT and KTH Vetenskap och Konst to evaluate emissions from print and digital media (Hohenthal *et al.* 2014). The study was based on the lifecycle impacts of both media by conducting a side-by-side comparison of three publications. The study found that one copy of the broadsheet size newspaper *Kauppalehti* would equal 224 gCO_2e (this equates to driving a car for 1.4 km). Reading the same content on an online platform is more complex. This is due to the conditions of content delivery as when reading a newspaper online, a reader uses electricity on the device they read as well as from the data server they access. The impact of online media increases with the time spent on viewing content, or on how many times a text is accessed or downloaded, whereas consumption of 'paper' for print media is more or less static in terms of the emissions from reading activities. They conclude that the longer one spends on reading content, the better an option print is: 'the majority of environmental effects of online media are caused by the actual reading' (Alma Media 2015), they conclude.

From the perspective of a single user, then, reading content in digital form may be more harmful than buying a print copy. From the perspective of internal environmental policy, Alma argues that print is more harmful at the production stage as this is where it requires most of its resources. It mitigates these by having the paper pulp either PEFC or FSC certified – international certificates that guarantee that pulp was sourced from forests that were managed in a social, economically and ecological manner. While the production of online media is arguably less harmful to the environment, it comes with many caveats due to the impact the materials used for the production of the terminal devices has on the eutrophication of waterways, mineral depletion and toxicity, among others. While online consumption may be more harmful than print, comparison between them is difficult due to a range of variables often outside the control of the company. We would need to pay attention to the reading habits of individuals and the specificities of the device, including its premediatic materialities and afterlives, to be able to contrast one against the other. Currently, for a company like Alma, it is only feasible to control emissions on the production side rather than in consumption. This extends to presenting statistics on user footprints: 'An annual subscription to the print *Aamulehti* corresponds to a maximum of 1.3% of the annual environmental impacts of an average European customer, while reading Aamulehti.fi for one year is responsible for only 0.6%' (Alma Media 2017a). The lower total for online is calculated on the basis of usership as emissions for online performance are lower due to the smaller number of visitors to the Aamulehti.fi service when compared with the total of papers printed – but again, these would be different for single users, and the different audience demographics of specific publications. Thus, carbon emissions need to be evaluated from multiple angles to come to an understanding of the impact of digital and 'traditional' consumption and production methods.

At the same time, these strategies provide opportunities for image management and branding, as a video clip on the Alma website concludes by stating that evaluating the footprint of different production and consumption methods 'transforms a risk into an opportunity'. While branding is often perceived critically in environmental writing, often for a good reason, framing sustainability as an opportunity is a tried and tested approach, suggesting that environmental policy can be a means of competition and, significantly, of being ahead of the curve. In comparison to broadcasting, the core operations of publishing have a considerable footprint which the company addresses even in its code of conduct. These areas are crucial as the publishing and print industry are responsible for an estimated 0.54 per cent of Finland's climate emissions. Adoption of these practices gives Alma a firm footing in the markets as the environmental protocols and practices for digital media it develops are transferable for the sector as a whole. Consequently, sustainability is best conceived as a form of reputational capital, where opportunities come from developing good practice in the organization and then using these as part of its industry steering practices. Here, Alma has made ample use of sustainability to promote its brand while advancing sustainability management beyond the responsibility deficit of other sectors.

Sustainability policy for digital media

Digital is omnipresent in all contemporary media, but it is a particularly central concern for publishing due to the material base of its core competencies. It has been framed as an opportunity for the sector to develop, but in order to do so, more comprehensive understanding of the energy resources of digital infrastructures are required. Digital communications pose considerable challenges for evaluating the scope emissions of publishing companies that need to consider the total supply and value chain of their operations well beyond their internal management practices. As Nicole Starosielski has suggested in relation to the digital footprint, 'internet infrastructure is dependent upon the mining of minerals, the unsustainable labor practices in technology production, and the energy-intensive geographies of data centers'. Drawing on the work of Lisa Parks, she argues that 'networks are powered by an array of electrical systems, maintained by vastly different laboring bodies, and are entangled with diverse social and ecological practices' (Starosielski 2016: 41). These are the complications facing companies such as Alma and Schibsted in that it is not sufficient to consider only their supply chain and internal operations but they have to start including these 'pipeline ecologies' into their emissions evaluations. Currently, the industry is not able to account for all the scope 2 and 3 emissions of the sector, even though strategies are in development.

A likely complication for these initiatives comes from what to include within these measurements. Alma has, for example, paid a lot of attention to devices and reading time, but other considerations, such as the use of files and access to cloud services provide specific challenges. Shane Brennan's work on backup culture (2016) touches on this topic in relation to digital. The use of cloud services to provide endless backups is seen and marketed as a way to ensure one's data is preserved indefinitely against disruption. They come with the promise of immateriality and operate as a potent means of data protection and convenience. But by multiplying the amounts of a copy in the server as well as providing more connected hotspots from where to access it, they require a considerable increase in the use of energy (ibid.). Certainly, they provide for efficient corporate conduct and management of information, but they are also quintessential examples of anthropocentric logic. The image of the immaterial cloud floating somewhere out there ignores the grounded realities of the data centres, still often at least partially powered by coal. Brennan perceives this as a contradiction in terms as cloud technologies and backup culture contribute substantially to the very risks from which they are nominally trying to shield us:

> By contributing in significant ways to climate change, ironically, the proliferation of cloud backup is helping to create more severe storms, floods, and power outages – the very things from which it strives to protect data. As these risks become more immediately visible, digital backup practices respond by creating even more redundancy, constituting a feedback loop.
>
> (ibid.: 58)

And yet, the dematerializing rhetoric of the cloud, as well as the use of 'sustainable data' to describe the recommendations for this practice for companies relying on it for their lifeline, have worked to culturally obfuscate this pressing environmental media circuit (ibid.: 58).

The proliferation and pervasiveness of digital media have huge costs which extend to the ways the 'dematerializing rhetoric' of the use of cloud services obfuscates most of the discussion of the environmental footprint of digital media. Alongside the contribution of planned obsolescence of media devices, research conducted by Elsevier, Alma, VTT and KHT, among others, makes it clear that the materiality of digital media needs to be challenged at the organizational level. There are certainly benefits to digital in comparison to the material waste generated by print but the changing practices of the publishing industry make it urgent to increasingly address their environmental role. Research is often commissioned on the topic, such as was done by Alma above, but outcomes invariably emphasize uncertainty as answers rely on too many factors and variables including reader habits, data farms, internet service providers, device manufacturers, and media companies. In principle, both Scope 1 and 2 emissions are well covered by publishing companies, but Scope 3 emissions are much more complicated when assigning responsibility, which is a problem considering these take up the large majority of their emissions. Most media companies do not own data centres as they outsource processing and hire storage facilities, but to understand the total emissions of a company, these would need to be factored into their accounting. For example, the dissemination of content produced by a company like Alma on webservers would require the distributor to provide specific data on usage patterns. Finally, data collection on user habits in terms of both the devices they use and the time they spend on content would have to be assembled through comprehensive access to user behaviour. Consequently, the problems for the industry as a whole is to do with not only agreeing on similar notions of transparency and common standards of accounting, but also collecting data from all these diverse sources.

Digital media and sustainability: Schibsted

The Swedish publisher Schibsted has also taken both the challenges and advantage of digital on board in its repositioning of its brand and its core competencies. The company has an active CRS policy (Schibsted 2017a) which includes both green reporting and elaboration on how it can influence the environmental brainprint of its consumers. It plays a role in the Responsible Media CSR Forum alongside Alma and has consolidated wide-ranging environmental protocols including a 'Green House' as its HQ in Stockholm. This enables the company to collate its domestic activities in Sweden in one energy-efficient building and use the building as a catalyst to reduce its footprint by a half. All companies that form a part of the larger Schibsted Group have agreed to operate according to the environmental regulations set by the central management. The company has also required 20 of its largest subsidiaries to provide data for the Carbon Disclosure Project. These include elaborate measures on handling chemicals used in printing to having waste processed by professional transport companies. They negotiate with pulp and paper materials suppliers and use the Nordic Swan Eco Label and work with consultancies like Grönt Punkt to address waste from unused newspapers, cardboard, waste paper and waste products from paper reels. Furthermore, they argue that the move to digital has reduced their emissions by 34 per cent from 2010 to 2014.

This provides an interesting benchmark for the industry as a whole but does not tell us much about how a company such as this rebrands its operations and how digital is changing its operating parameters. Much as with Alma, Schibsted is focused on 'transforming from a media company to a technology company' (Schibsted 2017b). This shift in the operational focus means that the environmental views of the company will also have to be renewed. Most of these strategies are communicated through their publication, *Future Report*, that forecasts key areas of its strategic development. Several articles in the publication focus on new areas of innovation that have potential implications for the environmental performance of the company, in addition to rebranding it as a service provider. While these do not directly address managing their environmental footprint, they do indicate the ways Schibsted tries to stay on top of the *Zeitgeist* by including, for example, discussion of the food consumption habits of millennials, which, according to the company, will require apps to capitalize on how sustainability is incorporated into these habits. Most of the document focuses on key flashpoints like AI and virtual reality, but a particularly significant area for environmental concerns comes from the various marketplaces Schibsted either hosts or includes among its sub-companies. Ownership or collaboration with platforms like Leboncoin in France, Blocket in Sweden, Vibbo in Spain, Finn in Norway and Subito in Italy provides business benefits, which also generate emissions savings which 'add up to 12.5 million tons CO_2e! This huge amount of greenhouse gas is the equivalent of flying an Airbus 380 around the globe 1,100 times. Or putting a total standstill of the traffic in Paris for three years and four months' (Schibsted 2016: 3). Yet, even as these reductions are impressive, they are based on creative rhetoric and accounting as the savings are premised on activities that arguably extend even outside of scope 3 emissions: 'The method behind the calculation is based on the assumption that each sold used product replaces the production of a new, equivalent one including the waste management of the product' (ibid.: 7).

While framing the hosting of market platforms as sustainable activity can be considered problematic, the ways the company integrates these into its core competencies are more appropriate. Schibsted provides the digital service Kundkraft to allow users to pool together to negotiate a new contract prioritizing green resources at their electricity auction. This gets to the essence of how the Schibsted brand operates. They are no longer a publishing house but a multi-platform service provider basing their business models on convergence and user experience. By tapping into the sharing economy and the expectations and user patterns of digital natives, the company uses these strategies to capitalize on the wider social significance and visibility of sustainability. This is most evident with a venture focusing on the use of digital platforms to enhance recycling, *The Second Hand Effect*. This venture ties together community building, user experience, platform provision, interactivity, and sustainability:

The Second Hand Effect is more than just numbers. For many people it's a social movement and a way of life. It means meeting new people, embracing a passion and being part of a community, trading things from one hand to another. It is an individual choice to live a more sustainable life.

While most of these examples appear somewhat unrelated to the sectorial model used in this book, they indicate strategic manoeuvres that publishing companies use in repositioning themselves as service providers. While in broadcasting, the debates tend to focus on meeting the expectations of a public organization, in the private publishing sector, innovation is key for these companies to meet challenges to their business model as well as to rebrand themselves. For the most part, environmental sustainability is framed as a PR opportunity that can aid in the rebranding of the company. Business opportunities that arise out of some of these recycling platforms provide both companies with a role in the sharing economy but they also tap into and encapsulate the new business models that Schibsted and Alma are pushing. Innovation is seen as a strategic concern that is essential for these companies to succeed in the changing marketplace – and to be able to do this credibly, the companies have to meet strict environmental standards. The move into these markets, combined with the visible footprint of paper as a symbolic reminder of the environmental impact of their traditional core competencies, has arguably led to the adoption of stringent environmental management measures.

A culture of sustainability

Environmental sustainability for many of these publishing companies is thus emerging as both a strategic and material concern that can affect their profits and operational conduct. These are pragmatic realities of operating in the contemporary multimedia marketplace, but a lot of their work is also focusing on moving from assessment to action (or from only reporting on environmental impacts to doing something about them). The company Shaping Environmental Action, providing consultancy and strategic services for businesses interested in environmental LCA, suggests that this is where the next steps in environmental corporate activity will be. In an article for the industry sustainability publication ISO Focus (2014) outlining the latest developments in ISOs, they suggest that reliance on Life Cycle Assessments (LCAs) and Environmental Management Systems (EMAS) may not be enough. These are a set of management and leadership tools for setting up environmental goals and produce data to aid in preparing for any transformations in legal structures. The problem is that they have not been fully implemented in any of the organizations that operate in media (partially due to the complications of responding to Scope 3 emissions). And when mechanisms are in place or under development, their application can be vague and indecisive, as we saw with some of the public service broadcasters.

Shaping Environmental Action argues that the problem is not only with know-how but also the culture of environmental responsibility:

> Our experience and research show that neither environmental management systems nor LCAs addresses the issues businesses face that go beyond technical difficulties. These barriers relate to business strategy, corporate structure, decision-making processes, information management, corporate culture and employee performance management.
>
> (ibid.: 15)

While both broadcasting and publishing tend to focus on infrastructural and technical problems, the increased emphasis on corporate culture with Alma and Schibsted suggests a shifting focus in organizational management. If the environmental sustainability KPIs of both sectors are concerned with meeting legal requirements and financial imperatives through technological processes that LCA and EMAS evaluate, they also include a heavy emphasis on office and staff management. In many of these companies, especially prominently with YLE and Alma, it is not only technical issues that are significant, but also identifying specific areas of responsibility for different roles within the company and integrating this philosophy also into the external contacts of the organization. They, as well as the shifting focus of environmental sustainability from operational to strategic or material concerns, may allow these companies to facilitate the creation of a culture of environmental responsibility. While this may not be up to the standards of the ecological habitus discussed earlier, especially as many of the companies are still finding their feet in terms of integrating sustainability on all levels of management, they can act as important indicators in transforming the everyday operating priorities of these companies.

Bibliography

Alma Media. 2014. *Annual Review*. Helsinki: Alma Media.
Alma Media. 2015. 'What is more ecological: print or online media?' Available at: https://vimeo.com/54275307 (accessed 11 November 2017).
Alma Media. 2017a. 'Mission, vision and values'. Available at: www.almamedia.fi/en/about-us/mission-vision-and-values (accessed 11 November 2017).
Alma Media. 2017b. 'Sustainable Alma Media'. Available at: www.almamedia.fi/en/about-us/sustainability/sustainable-alma-media (accessed 11 November 2017).
Alma Media. 2017c. *Ympäristö*. Helsinki: Alma Media.
Brennan, Shane. 2016. 'Making data sustainable: back-up culture and risk perception', in Walker, Janet and Starosielski, Nicole (eds) *Sustainable Media*. New York: Routledge, pp. 56–77.
Eurosif. 2012. *Media Sector*. Brussels: Eurosif.
Hohenthal, C., Moberg, Å., Arushanyan, Y., Ovaskainen, M., Nors, M. and Koskimäki, A. 2014. *Environmental Performance of Alma Media's Online and Print Products*. Tampere: VTT.

ISOFocus. 2014. 'Green growth', *ISO Focus*, July–August 2014.

Kennon, N., Howden, P. and Hartley, M. 2009, 'Who really matters?: A stakeholder analysis tool', *Extension Farming Systems Journal*, 5(2), pp. 9–17.

Masalin, Heikki. 2015. *Mapping of Nordic Creative and Cultural Industries: Financial Enterprise*. Copenhagen: Nordic Council of Ministers.

Moor, Liz. 2007. *The Rise of Brands*. New York: Bloomsbury.

Rhenman, Eric and Stymne, Bengt. 1965. *Företagsledning i en föränderlig värld*. Helsinki: Bonnier.

Schibsted. 2014. 'Vision and values'. Available at: www.schibsted.com/en/About-Schibsted/Vision-and-values/ (accessed 11 November 2017).

Schibsted. 2016. *The Second Hand Effect*. Oslo: Schibsted.

Schibsted. 2017a. 'Environmental footprint'. Available at: http://howwecare.schibs ted.com/environment/ (accessed 11 November 2017).

Schibsted. 2017b. *Future Report*. Available at: https://futurereport.schibsted.com/tech nology-empowering-people/ (accessed 11 November 2017).

SFS. 2005. *The Swedish Companies Act*. Available at: http://law.au.dk/fileadmin/www. asb.dk/omasb/institutter/erhvervsjuridiskinstitut-skjultforgoogle/EMCA/Nationa lCompaniesActsMemberStates/Sweden/THE_SWEDISH_COMPANIES_ACT.pdf

Starosielski, Nicole. 2016. 'Pipeline ecologies: rural entanglements of fibre optic cables', in Walker, Janet and Starosielski, Nicole (eds) *Sustainable Media*. New York: Routledge, pp. 38–55.

Strand, Robert and Freeman, R. 2015. 'Scandinavian cooperative advantage: the theory and practice of stakeholder engagement in Scandinavia', *Journal of Business Ethics*. 127(1): 65–85.

9   Film and television

Introduction

> Hollywood's green conscience to this point is a marketplace conscience, torn in conflicting directions by the forces of economics, industry, and public image. Recognizing the economic benefits of sustainable practice, studios have begun to tighten the efficiency and renewability of their raw material use, foregoing, however, any radical industrial change or empirical critique of the price our planet pays for this culture of excess. As such, the discursive machinery of Hollywood manages to deflect popular criticism of the environmental ramifications of film production, and to avoid governmental regulation of its practices.
>
> (Vaughan 2016: 27)

The exploration of both the broadcasting and publishing sectors has underlined the notion that environmental sustainability for the media is always a complex compendium of ethical, financial, organizational and legislative intentions and motivations. There is no simple solution – whether managerial, regulatory, financial or ideological – that would work for all the different sectors. For the most part, sustainability poses an obligation for broadcasting, whereas it has largely been framed as an opportunity for publishing companies. Its relationship with film production is even more complex. This is a sector that often relies on private investment and focuses its fortunes on a restricted number of very expensive products, at least in the Hollywood business model. Vaughan's arguments touch on key ways Hollywood's attempts to promote green production methods and environmental literacy are fraught with contradictions. Highlighting environmental content in resource-intensive texts like *Avatar* (2009) or valorizing the role of superstars like Leonardo DiCaprio in promoting environmental awareness masks beneath what is effectively an extensive, and expensive, PR operation designed to capitalize on the promotional or branding value of a film or its star. This is green capitalism at its most efficient, where relatively small environmental advances are reciprocated by large returns on investment.

The Nordic film industries have a very different relationship with regulatory infrastructures and public and private capital investment. As these are small national cinemas (Hjort and Petrie 2007), they rarely reach the levels of financial success of Hollywood cinema or, most accurately, even gain solvency through market competition with, predominantly, imported products. They thus have to rely on government support to maintain the operational capacity of the industry and continue to facilitate productions. Some larger companies like the Danish Zentropa or Finland's Solar Films have the international connections and domestic audience share to maintain consistent operations. Yet, the majority of companies tend to be small or medium-sized operations with an output of four or five feature films per year. Limiting their reach is the size of the domestic audiences necessitating that practically all production is government-supported through national film institutes. Additionally, much of this production relies on national broadcasting and television companies for financial support with rights sold in advance across the Nordic countries.

The resulting circumstances for environmental sustainability are complex as the private ownership and size of these companies preclude them from having to abide by many of the larger regulatory frameworks, such as the environmental quality objectives (EQOs) in Sweden or the Principality Decisions in Finland. At the same time, the regulatory infrastructures of the state govern general laws and policy of areas, such as office operations which influence the day-to-day management of emissions. Furthermore, if film companies are commissioned to produce a work for YLE in Finland or SVT in Sweden, they will have to abide and report on their emissions, as both companies have started requesting reporting from their collaborators. But as the production of theatrically released films is based on a piecemeal structure of public and private funds, enforcement of these protocols – for procurement of production technology or for works that are only partially supported by YLE or SVT – tends to be limited on productions with a minority stake from public funding. At the time of writing, individual production companies do not have consolidated environmental strategies in place. The consequence is that any full-scale adoption of sustainable production practice would need to be dictated on the level of institutional policy or through funding mechanisms necessitating carbon reporting. Problematically, none of the Nordic film institutes have received mandates to develop environmental policy (partially due to the responsibility deficit outlined above), and consequently there is no independent oversight of the industry's environmental role or established initiatives to develop means of self-regulation.

Film production has only recently seen funding organizations start to develop initiatives to address these issues. These organizations are small regional film commissions in Sweden as nothing exists as yet at the national level. In Southern Sweden, the regional film fund, Film i Skåne/Öresund Film Commission, and the production company, Filmlance, participated in the 'Sustainability in Vision' study, commissioned by the European network of film funds, Cine-regio (Cine-regio 2014). The resulting study component based in Sweden focused on the television show *The Bridge/Bron* (Season 2) and resulted in a range of suggestions for implementing environmental protocols on set. These were based on hiring a Green Runner to oversee the administration of a crew management plan for the duration of the production. Other areas included the creation of a training programme for local industry, a Green Crew memo to be distributed at the beginning of production, Green Procurement and product placement campaigns, sustainability tracking and assessments that would benefit future productions, and energy- and fuel-reduction campaigns that engage the cast and crew in order to gain their support. Furthermore, the plans emphasized the need for time to be allocated at the beginning of the production process to raise awareness of sustainability among the crew as well as a transparent recycling system that 'would sustain the crew morale and would also have an environmental impact' (ibid.: 30).

Film I Skåne is thus contributing to a viable framework to change the approach to sustainability in the Nordic countries but this is still at a very early stage, and is only designed, currently, to operate at a regional level. Simultaneously, Film i Väst, another regional fund, is working on sustainable modelling with Chalmers Technological University, with the aim that these will eventually filter into national policy. The encouraging part of these operations is that they address most of the pressing infrastructural concerns on integrating sustainability into film production practice. Most importantly, they provide appropriate practical guidance designed to meet the core competencies of the industry. Clearly, there is some burgeoning interest in these themes, but these are at such an early stage that they do not provide much policy material for analysis here. Instead, we now focus on key themes emerging from the Nordic film and television production context that draw on both medium specificity and the significance of the political and cultural context of the industry. These strategies illustrate context-specific means of merging environmental activity with strategies that, arguably, are suitable for the context of Nordic film and television production. In doing so, they emphasize that environmental policy should be shaped not only for the requirements of each sector, but also for the social and cultural context in which they are formulated.

Infrastructure

Infrastructure is often used as an explanation for the absence of comprehensive policy on sustainability for the media sector. What we have called the responsibility deficit – the argument that Denmark, for example, aims to be carbon-neutral by 2050, and media will be swept along – has resulted in several complications for the sector. The broadcasting and publishing companies analysed here have had to respond in ways that both meet any (often general) regulatory directives imposed on them and accommodate the specific requirements of their core competencies. The circumstances of film and television production, again, imply a distinct set of requirements as they struggle to identify their responsibilities and potential agents that could solidify action on sustainability. To reflect this lack of industry coordination, the majority of attempts by me to contact individual production companies and institutes resulted in a polite decline as these organizations had not considered or addressed these concerns in any sufficient depth. At the same time, several producers and cultural authorities were willing to discuss their views on these matters. The following explains some of the key arguments presented by these individuals and organizations in terms of how they relate to wider debates on environmental sustainability.

The media in Iceland have arguably been overlooked in this book due to a complex set of circumstances where the Icelandic media industry is both comparatively minor in size and supported by an extensive green infrastructure. These circumstances mean that the publishing sector is both catering for a small population of 332,000 people and relying on a largely indigenous resource base to do so. These concerns apply to other sectors as well, as the small scale of the operations and the country's electricity mix arguably make media production sustainable. This amounts to a highly advanced version of the responsibility deficit that suggests Iceland is already meeting global environmental objectives and, by extension, all media production is already environmentally sustainable. This is the view of the Film Commissioner of Iceland, Einar Tomasson (pers. interview, February 2016) who suggests that the country's infrastructure facilitates greener productions than in other comparative cases. When asked if Iceland would need environmental media regulation, the answer is clear: 'We do not need calculators as 99 per cent of the heat for our buildings comes from thermal power and the only fossil fuels we use are for our car fleets' (ibid.). The comparison is especially pointed at the UK, where, arguably, environmental policy for film and television is more advanced, but the reliance of its energy mix on coal and nuclear power makes it much more unsustainable than Iceland by default. As Iceland is able to tap a vast range of renewable reserves, it is able to provide the media with a means of production that is inherently sustainable, Tomasson argues.

Renewable energy is a key asset in providing weight for the infrastructural argument. Tomasson argues that it underlines most areas of the production cycle including storing products and data. Server centres run on renewables that use the outside air for cooling every day of the year. According to him, 'Nordic power is interconnected so electricity is pumped into the grid no matter if it's hydro, coal or nuclear. One has no idea where the electricity is coming from. But here in Iceland it is only from one source' (ibid.). The energy mix complicates claims of a country like Sweden as their energy infrastructure is heavily reliant on nuclear power. The same goes for the UK as it uses an undefined mix through its grid. For him, this is a problem to do with imposing any form of regulatory oversight to cover Europe or even the media as a whole: 'The use of one approach fits all does not in fact fit all and has to be modified to take into account contextual differences' (ibid.).

This makes sense but also indicates that the responsibility would have to be borne by domestic organizations in charge of cultural and economic policy. But simultaneously, he suggests that 'arguing that a country is not doing enough can be problematic when the whole country is environmental' (ibid.). This is, of course, the responsibility deficit in action as Tomasson makes the point that specific legislation or policy would not be required due to infrastructural concerns. While this argument makes sense, simultaneously, the particularities of film production arrangements in Iceland raise pertinent questions over how much of the total range of film production in Iceland can be considered environmentally sustainable. For one, the reliance of the Icelandic film economy on runaway productions from Hollywood brings in key elements that complicate any isolationist view of the Icelandic energy infrastructure. Due to tax breaks for productions and the unique landscape of the island, it has acted as the host of shoots for several huge franchises, including *Game of Thrones* (2010–) and *Batman Begins* (2005). For Tomasson, these productions have to abide by heavy regulations on the environment. Productions have to work with the Icelandic environmental agency as well as a production service company such as True North who advise the US productions on what can and cannot be done on locations. The idea is to have nature in a 'better situation after they leave' (ibid.) instead of them polluting or using its resources unsustainably. These productions will also need to adhere to mandates from the PGA or policies imposed on a studio level – as would take place on Twentieth Century Fox shoots, for example. Infrastructure also plays a role in travel and accommodation as hotels are heated by renewable power and batteries are 'probably' charged with renewable energy from the Icelandic grid.

Yet, this does leave open a lot of questions. The first level of concern is infrastructural. While these productions may charge equipment with renewables, we do not know much about the equipment itself. Nor do we get figures on how much power they consume when in Iceland and how this compares with domestic productions. Hollywood productions inevitably rely on large-scale travel, both inside and to Iceland. Air miles and the use of trucks to transport equipment make up a huge proportion of the carbon footprint of a film shoot. Electric cars are becoming increasingly commonplace but travel is not addressed fully for all its diverse requirements beyond electricity use. The argument to leave the environment in its original state is a worthy ambition but it also only skims the surface of the detrimental effects that large shoots comprised of potentially hundreds of crew members have. Recycling waste from food consumption or packaging is fine, but this does not adequately calculate the visible and invisible effects of film production. These include the technical specificities of imported technologies, the consumptive practices of the crew on and off set, the percentage of plastics in sets and other materials, or the use of intercontinental cable connections to communicate with the HQ. Simply the fact that the Icelandic film industry relies so heavily on these productions puts its environmental credibility in question as these arguments do not take into account a whole range of externalities that will influence the footprint of the industry as a whole. In light of all these concerns, the evocation of the infrastructural argument does not seem fully sustainable in its own right.

Locality

The comparatively small scale of Nordic film productions leads to another pertinent argument on its environmental role – its emphasis on local shoots. In comparison to Hollywood productions, film shoots in the Nordic countries tend to be restricted by location and the funds available for production coordination. Consequently, crews are small and shoots are done on existing sets or locations. Eric Vogel, lead producer at Torden Films in Norway, explained in an interview in July 2015 how this relates to sustainable production in Oslo. The city is one of the key hubs for filmmaking in Norway as most of the domestic talent pool is located in the capital and most film and television shoots take place on location there. Vogel suggests that this enables local coordination of production in a way that is inevitably sustainable. First, concentration can productively limit the need to commute from outside of the capital. The second is the way the productions are coordinated within the city. Film production is largely a local operation and thus travel, sustenance and accommodation are coordinated with local resources that already often have sustainable procurement practices due to infrastructural municipal or governmental regulations. These extend to all parts of the shoot from the filming equipment to food and water and provide ingredients for a sustainable film industry. In these situations, saving on car journeys, for example, by using car pools or public transport would actually make sense from a production management perspective.

The argument is not that different from the responsibility deficit cases, though Vogel suggests that creatives play a key role in these arrangements. They tend to be environmentally conscious in their daily life which translates to added pressure for film production to go sustainable. These also apply to distance travel if talent comes from other parts of Norway, as flights tend to be cheaper than trains. The means of travel are an ethical choice that often has to be left to the individual. But at least, according to Vogel, creatives tend to consider environmental sustainability as an integral part of their work practices. Whether this translates into a form of ethics or acts as publicity management is open to debate. Most importantly, in contrast to the large mobilization of forces involved in a Hollywood production, small industries work with limited budgets leading to a situation where smallness can save on aspects of the footprint (by the necessity of finding local options or cutting back extravagant costs), but they can also restrict what can be done to curtail the environment impacts of film production. There are no means to hire eco-supervisors nor are there clear directives or frameworks in place for instigating sustainable production within the Norwegian media industry. Instead, this is very much community-led activity where it is up to individual producers to act locally and make purchasing choices that reflect their ethical values. Thus, the scale of production contributes an intriguing addition to the discussion, but also strengthens the sense that any tangible attempt to integrate sustainability into the industry will often be voluntary.

The role of film institutes

A more comprehensive approach to implementing environmental management structures to Nordic production would need to work on an institutional level of responsibility, as it does in many other contexts. The problem is that, at least at the time of writing, the film institutes in these countries have no policies in place for sustainability management. The Norwegian Film Institute's response to a query on environmental sustainability explains this situation well:

> The Norwegian Film Institute has no mandate or any authority to monitor or regulate these issues. In practice, the film industry is not a very environmental friendly industry. A production often requires transportation of people and equipment from place to place regardless of what impact this may cause. A film production is also very expensive and time is often crucial for keeping the costs as low as possible – How the crew operate and the choices they make are not always the most environmental friendly solutions. The priorities are based on getting the best results. Having said that, there are many in this business with a strong engagement for these issues, but there are no formal practice or group that deals with it.
>
> (Mette Tharaldsen, pers. comm., 2015)

The Finnish Film Foundation also expresses interest in these developments but has not started to implement any concrete means to either adapt or develop their own versions of these policies. This is a problem to do with responsibility to a large extent, as these institutes indicate they would expect instruction from a ministerial level. Unfortunately, as we have already chronicled, the cultural or the environmental ministries in Finland do not consider it their role to legislate for the media or see this a problem of sufficient scale to require their attention. The media are often perceived as a vital organ of the democratic state and curtailing its operations even in terms of production structures has been met with hesitation due to various notions of impartiality. At the same time, the cases of BAFTA and the BBC show how environmental imperatives can be developed from voluntary measures to mandatory requirements with success. The reality of industry adoption of these protocols is, as most of the organizations frequently make clear, a complex and multifaceted process, requiring a range of techniques from incentives to regulation.

The regional film funding organization, Copenhagen Film Fund (CFF), provides a good example of how different regulatory regimes and cultural and financial incentives contribute to developing sustainability policies for film production. Thomas Gammeltoft and Caroline Gjerulff, film commissioners at CFF, view film production's footprint as an 'obvious area' of focus and an issue 'we all have to address' (pers. interview, 2016). As with the Norwegian Film Institute, CFF notes that there is a fundamental contradiction in the business 'as we are heavy carbon users in all we do'. The CFF does not have a strategy in place yet but has been networking with European organizations to develop its own strategies. The first steps they have taken have been to focus on general operational areas like cutting paper use and necessitating recycling. These consist of 'simple regulations or restrictions you put on yourself'. Yet, they argue there needs to be more concrete institutional support at a sectoral level where the role that the Danish Film Institute (DFI) would have to play to establish and consolidate policies to make sustainability strategies part of normal practice across the industry would be central. Unfortunately, the DFI has not put any real plans in motion on this theme (at least at the time this interview was conducted).

Infrastructural arguments are still a part of CFF's strategies as 'minimum standards in Denmark are already high', but the role of film funds as public organizations necessitates that they have to be role models for what they preach. CFF identifies its role as helping productions by providing them with green alternatives and suggestions on sustainability. Another means of initiating industry interest would be to provide financial support for adopting sustainable reporting. Here, CFF uses the economic repertoire to overcome concerns that these strategies would mean increased costs for shoots. They argue that not only will green methods be subsidised to alleviate expenses for early adoption but they would potentially cut costs in the long run. Financial incentives such as offering an additional €100.000 when a production confirms it has minimized its emissions are considered potential strategies while also merging the cost-effective mentality of producers with creative use of resources, whereby adapting environmental strategies 'may cost the production DKK20,000 but can result in a return of DKK10,000 of funds' (ibid.). Such approaches are important as 'producers have so many obligations to adhere to in terms of financing that to ask them to add another obligation would be quite problematic'. An additional strategy would be to use eco-supervisors on set who would take care of recycling and limiting the footprint of the production. A problem would be budgeting, though the DFF suggests this would need to be compensated by institutional support and know-how through training. Considering these mechanisms do not currently exist in Denmark, imposing mandatory regulation on the industry would be difficult and also potentially detrimental to the adoption of these strategies.

DFF argues that these measures would need to be used strategically to generate awareness of the footprint of the industry where repetition of sustainability strategies would consolidate them as normal industry practice. They argue that 'the more we put it in talks and the more we direct it, the more we can do something about [the environment]' (ibid.). But even if this was achieved, it would have to be done on the basis of an extensive collaborative network that spans, and consolidates, the industry. This is especially a concern for small film cultures where individual actions would not amount to anything systematic or impactful. As was the case with the local industries of Norway, these would be more akin to lifestyle choices made by different creatives instead of a normalized protocol for film production. They would, ultimately, lead to conditions not much different from the responsibility deficit. Instead, a more focused consolidation of efforts on an institutional level is required. But simultaneously, this would need a wider level of international connectivity: 'we are not able to do this alone as our industry is more and more globalized. One isolated country does not state they can go CO_2-free' (ibid.). Consequently, the industry as a whole would require a concentrated cross-border effort to ensure the political ecology of European film and television production meets these demands.

A model for the sustainability network: the European cases

The film industries of the Nordic countries show a relatively low level of engagement with sustainability initiatives. Yet, the picture is very different in other parts of Europe. A range of consulting and financing organizations have established policies to integrate sustainability into European film production. Cine-regio is a network of 44 film funds that acts as a key organization coordinating between different regions to facilitate knowledge transfer and networking coordination. They provide a platform for meetings to discuss the latest developments in organizational strategy as well as publishing annual overviews of sustainability in the sector. This organization is a key obligatory passage point (OPP) for the network, acting as a central mechanism through which different regional organizations can collaborate. One of the key mechanisms facilitating this collaboration is the creation of various carbon calculators. In discussion with these organizations, the PGA Green Shoot guidelines are often seen as an inspiration while the Albert calculator by BAFTA is credited for advancing the European film and television sectors, especially as it pays attention to the restricted scale of production in comparison to Hollywood. Others include e-Mission by the Flanders Audiovisual Fund (which requires green reporting to receive the final part of project funding), Ecoclap by the Paris-based Ecoprod, and the Green Shooting Card by the Hamburg Schleswig-Holstein Filmförderung.

The head of sustainability at Ecoclap, a French organization overseeing environmental sustainability, Olivier-René Veillon lays out the opportunities in an optimistic manner: 'We have the tools. We have the financing. Now, for the companies, it's simply a matter of strategy and administration' (Green Film Shooting 2015a). The level of confidence in this statement is encouraging but the integration of sustainability into European film and television production is compromised by a fragmented approach and lack of comprehensive central coordination. Cine-regio certainly has worked on this front to build infrastructural capacity to ensure these considerations are addressed, but full-scale implementation of environmental sustainability remains largely an aspiration due to the slow pace of development in encouraging the necessary institutions to commit to providing funding for these initiatives. While there is certainly interest in these themes, they are not either consolidated by long-term investment or by sufficient human resource management to assure continuity. The following sections will focus on unpacking these problems – as well as crediting many of the advances in policy and practice – by elaborating on four types of management concerns: human resources, finance, regulation, and organizational networking.

Labour management

As many of the organizations addressed above emphasize, sustainability management works best when it is built on individual responsibility. Here, the clear allocation of managerial roles and responsibilities allows organizations to avoid an internal responsibility deficit and ensure that the specific emissions from production practices are addressed. For example, the Wallonian Film Fund in Belgium discusses the role of the director who should oversee all aspects of sustainability. These would not only involve artistic choices like LED lighting but also areas like transportation and the establishment of criteria for selecting hotels and catering services. While these areas would seem to be part of the roles reserved for production managers or producers, 'ultimately, all of these decisions should be discussed with the Director' (Cine-regio 2014: 5), they recommend.

There is ample evidence to support this assertion. Certain independent film producers have adapted environmental practice on their sets out of personal interest and without any push from an institutional level. German director Lars Jessen, for example, has implemented environmental practice on all his film shoots (pers. interview, April 2017). These include a strict code of conduct on recycling, the use of LED lights, and so on. He communicates these ideas to the crew at the outset of the production and often tends to work with the same individuals who are already familiar with his working habits. If these protocols are established as the norm from the beginning, and they are communicated as matters-of-fact for the production, there tend to be relatively few problems with individuals who find these practices cumbersome or who view them as an obstacle to an effective shoot. Jessen's responsibilities also range from discussing the potential to tap greener electricity to power the production, though these, he suggests, can be outside of his control if the shoot is in a location where access to green electricity from the grid is not feasible. The incorporation of even this extent of environmental management is only possible due to a commitment from an individual in a position to set these protocols.

Yet, these individual strategies are far from establishing an ecological habitus for the industry. For this to be consolidated, intensive management of production activities is required. BAFTA takes this view as well as 'senior staff should discuss and agree on the show's sustainability goals at every stage. At first opportunity, tell your cast, crew and supply chain that this will be a sustainable production' (BAFTA 2014: 4). Once these protocols are decided, the producer will provide cost estimates and negotiate with line producers on their implementation: 'it often takes the entire length of a shoot to get the whole crew accustomed to the waste and water policies of a sustainable set'. An example from line producer Jules Hussey explains the significance of clarity and control: 'I informed everyone from the start that this was going to be a drama where the environment mattered. That no one was going to be surprised when they were going to be asked to do certain things' (ibid.:5). These are also reflected in the Checklist provided by BAFTA as it emphasizes nominating a person responsible for communicating sustainability on set. Ecoprod agrees with this as well, with Veillon arguing for the central coordinating role of the producer in charge of sustainability: 'It has to be a criterion to evaluate the producer as well as the production. A good producer has to deliver accurate information on the production's carbon impact and reduce it' (Green Film Shooting 2015b: 5). Thus, individual motivation and responsibility emerge as key strategies for setting policy and coordinating labour. Here, publication of these strategies in the BAFTA Albert and Green Film Shooting literature is essential for consolidating these individual ventures into a pattern that the industry can follow.

To take this further, Filmförderung Hamburg Schleswig-Holstein, a regional film fund based in northern Germany, argues for the need to make environmental sustainability a more formal arrangement through subsidies: 'To encourage producers to commit to sustainable approaches', they can apply for funds to get a designated ecosupervisor for the project. These individuals 'take responsibility for sustainable decisions and communicating them to the team' (Cine-regio 2014: 10). Similar strategies are already in place at BAFTA who run a programme to train green runners. These are often pitched as entry-level positions for individuals with an interest in sustainability and film and television production. For Schleswig-Holstein, the PGA case study of Emillie O'Brien's role for Sony Pictures is used to exemplify how such management strategies take place: 'I meet the department heads and I assess all of the practical/logistical concerns,' she argues (Green Film Shooting 2015b: 3). O'Brien considers her role to be most effective when it is a physical visible presence on set that will influence the crew as she can provide them with best practice tips on set and lead by example.

Yet, for these sorts of actions to become normalized as industry practice, shoots would have to be able to employ eco-supervisors on a consistent basis – which is what Schleswig-Holstein, CFF and others are exploring as a permanent option. This is reliant on the provision of funds, which at least in the European context, would require considerable support from institutions such as Schleswig-Holstein. For Hollywood film studios, it may be possible for a production to reach sustainable status if their parent company buys sufficient carbon credits or employs an eco-supervisor as part of the studio labour force. The majority of European films tend to be small productions with limited resources who would find such measures cost-prohibitive. In attempting to mitigate the obstacles to recruiting additional crew members to supervise on-set sustainability, Cine-regio suggests that 'the role of an ecosupervisor may pay for itself through savings on set' (Cine-regio 2014: 4). This would make sense once a solid infrastructure is in place for the industry as a whole, but currently, the practice of employing eco-supervisors would necessitate transformations in film policy as well as capital investment in integrating sustainability as part of the core operations of funding film production.

Financial management

Cost savings and financial management have been identified as key factors in 'selling' the need for sustainability to the industry. The documents by Cine-regio and Green Film Shooting are replete with comments such as the example of Brussels-based Accompagnement des Compétences Audio-visuelles which has 'highlighted the cost-saving benefits of green production methods' (ibid.: 7). It simply is not enough to use inspirational logic or motivational rhetoric but even the policies encouraged by these sustainability documents show that emphasizing the economic bottom line is a necessity. They must come attached to 'inspiring examples, such as "How to save money with green production" (ibid.: 7). It is clear that

> it's difficult to convince an ordinary production manager to consider anything but the bottom line, but at the same time, all it takes is one person above-the-line to run a production sustainably... as either a line producer or an Eco Supervisor can take the responsibility to help the cast and crew adopt sustainable practices.
>
> (Green Film Shooting 2015b: 3)

Emphasis must thus be on ensuring that the financial basis for appropriate staff responsibility allocations are in place.

Financial management of sustainable media concerns not only human resources but also material considerations. BAFTA's instructions for best practice explicitly frame them as reducing costs: 'ensure your team has access to recycling bins and encourage their use; re-using tapes rather than buying new ones every time can cut waste and save money' (BAFTA 2014: 7). Cost effectiveness is intriguingly presented as an additional benefit to environmental considerations, which positions it as common sense. Such a recourse to logic creates the seeds for the ecological habitus to form as it can effectively normalize behaviour in ways that are important for focusing attention on everyday work practices. BAFTA uses this rhetorical device repeatedly: 'Try to use people and services such as catering that are based near to your location.' Other suggestions concern the supply chain as 'requesting low-carbon vehicles from all suppliers' or 'choosing locations that require minimal travel' (ibid.: 10) cut down on the costs of a production. Some of these strategies are specific to the media: 'We regularly have costume sales and the profits feed back into our budgets. Any old costumes from the sales go back to charity' (ibid.: 14). Other ones are basic office protocol: 'We've got much better at turning off our computers and TVs at the end of the shift. It just takes a second and we know it makes sense' (ibid.: 14). While these are all presented as part of environmental policy, readers are asked to perceive them from an alternative perspective as a cost-saving mechanism that would be standard practice for any economically sensitive form of production. By positioning these strategies as part of the financial management of production, BAFTA frames the adoption of sustainability as something more than an imposed mechanism.

French public broadcaster, France Télévisions, whose environmental strategy aims to reduce carbon emissions by 8 per cent by 2020, summarizes this economic rationale well:

> Broadcasters can save money by reviewing production methods to reduce the amount of energy and resources used. Sustainable production is making programmes in a way which will have the minimum impact on the planet and the maximum benefit to the people and places involved in their creation. We made investments in our studios in order to reduce the carbon footprint. This, of course, is an economic advantage in the long run. The investment pays itself off and then it becomes a long-term benefit.
>
> (Cine-regio 2014: 7)

Economic logic is thus used from the start where sustainability planning emerges as a way to alleviate problems in human resource and material management. This, again, emphasizes the key role that policy must play in establishing an ecological habitus. Here, economic rhetoric helps to prepare the industry for the inevitable necessity of actively working with consultancies to lobby national film institutions and responsible cultural organizations to provide institutional funding and training provision to enhance their environmental sustainability. Anticipating legislative transformation may appear wishful thinking at this point in time, but there is also ample evidence to suggest that sustainability processes will become increasingly mandatory – as has been the case with the BBC, for example.

Thus, the legal and economic drivers, to quote Greening Film (2017), must work in tandem. Strategies such as these can be seen, for example, in the work of the Centre National du Cinéma in France (National Cinema Centre,

CNC) who are implementing a new funding system to support environmental actions (Green Film Shooting 2015b: 4). The CNC launched a new initiative in 2014 to cover half of the costs of productions investing in sustainability on set. The funds 'are not particularly large' as initially, for 2014, they supported one project with a €24,000 grant and for 2015, three projects with grants totalling €85,000. Baptiste Heynemann, Head of Technique and Innovation at CNC, suggests that these funds could be used for eco-supervisors, and need to address other physical needs like selecting biodiesel over other fuels, adopting hybrid cars, rechargeable batteries, recycling wardrobes, or prioritizing the use of LED lights. Clearly, then, the use of financial incentives is emerging not only as a rhetorical motivator but a concrete strategy.

Standards and regulation

Positioning sustainability tools as economically beneficial is a powerful means to entice producers to use them, but as we have suggested, organizations like the BBC have started to introduce obligatory measures into production arrangements. France Télévisions also considers obligatory reporting an essential part of this calculation: 'The next step is to enshrine eco-friendly production practices legally in the terms of coproduction contracts' (Cine-regio 2014: 7), they outline. Introducing mandatory mechanisms for reporting will have reverberations for the film industry as such measures are a heavily contested area of policy for the industry as well as the regulatory organizations. We have noted areas such as the responsibility deficit and the arm's length principle as instances where legislation on environmental sustainability has been challenged or obfuscated. But to facilitate the successful adoption of these practices, the positive value systems of environmentalist and economic repertoires are often not enough for full-scale implementation of sustainability – more thorough regulation is frankly necessary.

A case study of Dogwoof Films, a documentary film distribution company in the UK, conducted by the promotional site, Greening Film, illustrates this well. They suggest that 'fundamentally sustainability is an easy thing to grasp' (Greening Film 2014), which they aim to illustrate with Dogwoof's adaptation of these practices. The company has an interest in sustainability as they are 'the UK's leading distributor of ethical films'. It would make sense for such a company brand to incorporate sustainable production and distribution practice to align its ideological principles. This is especially important in the distribution environment where reputational areas will impact on consumer choice. Dogwoof ended up starting a trial and created an initial sustainability policy based on the principles of the sustainability standard BS 8909, which has environmental requirements on carbon footprint, waste management, biodiversity and economic implications on local investment and long-term viability.

The trial had some intriguing results as the 'Dogwoof objective is to … reduce carbon by 10% by 2015' (ibid.). Yet, the company was 'slightly overwhelmed' by the amount of work needed to fully implement BS 8909. They discovered that to implement the requirements of the Standard would require one person in a full-time role. As one of the aims of the Standard is to set out plans to demonstrate improved sustainability performance within available budgets, the complications faced by this relatively small organization to implement the strategies within budget are indicative of the wider problems pertaining to environmental sustainability strategies. To meet the full requirements of a thoroughly managed operation, much more comprehensive investment is required. Financial commitment from the organization or, in this case, from an institutional level, would thus be an essential requirement for these strategies to work effectively. The Chairman of the company, Andy Whittaker seems to corroborate this suggestion as he emphasizes 'the need for environmental and social responsibility [to be seen] as an integral part of the business model for a successful film industry' (ibid.). This suggests that instead of considering sustainability as an obstacle, it needs to be framed as an organic part of planning from the onset, as we have also argued. Yet, such rhetoric is fine in principle, and works in individual cases, but the transformations in the industry's contemporary habitus is nowhere near this level of commitment.

Thus, formulating strategies and establishing functional unilateral policy for film and television are besieged by the industrial necessities of making it part of the business model while trying to limit the economic and operational costs its implementation has. Yet, Ecoprod makes the case that this will immanently become part of a larger question to do with the general political will to go green, a suggestion that corresponds with many of the arguments made across the regulatory and industry spectrum on balancing imposition of regulation and individual motivation. The comment indicates the complexity of finding agreement on how best to implement environmental sustainability. In contrast to the ministerial response – often heavily reliant on the responsibility deficit – Ecoprod perceives the problem from its perspective as a cultural organization: 'We need to have the practical solutions in order to provide the political will the wherewithal to do everything that's possible. Of course, the political will is always important, especially in our business, which in France is heavily regulated' (Green Film Shooting 2015b: 7). For Ecoprod, a commitment from the regulators is key to ensuring the industry complies with its responsibilities. As receiving this level of investment from state governance may be difficult, the self-regulation model seems to be the most feasible approach to entice small to medium-sized production companies to cooperate. The chain of responsibility starts with organizations like France Télévisions and Ecoprod, or BAFTA and the BBC, as they can provide both adaptable strategies and financial incentives for the industry. Incentives such as the Carbon Literacy training at BAFTA and the necessity of reporting emissions for all internal and external co-productions for the BBC are functional models of policy that establish the required parameters for instilling sustainable behaviour as a normalized part of planning and behaviour on set.

Organizational networking

While management of resources, capital, staff and finances have an important role to play as consolidating mechanisms, communications management is vital in delivering these to their intended audiences. This is especially important as the film industry tends to use the environment as a PR tool, as Vaughan (2016) has argued. Here, 'to maximize visibility, the industry has cosmetically shifted the aesthetics of the star system, with websites such as Ecorazzi.com taking advantage of photo ops to show icons like Leonardo DiCaprio driving around in a Prius' (ibid.: 27). This sort of green capitalism is a concern but also arguably a pragmatic tool. Most areas where the brainprint is discussed feature a similar set of concerns that highlight the role of positive PR through flashy campaigns and the use of celebrities and other recognizable iconography. Intriguingly, these share an approach with the way consultancies within the film industry promote sustainability. These include the German Green Film Shooting (GFS), arguably the leading organization for Europe, a collaboration between the journalist Birgit Heidsiek and the Filmförderung Hamburg Schleswig-Holstein. They produced the *Eco-Cop* publication in 2015 in collaboration with the World Bank's Film4Climate initiative and Ecoprod, to coincide with COP21, which featured a host of articles focusing on technicians in the industry, policy-makers, producers and celebrity environmental ambassadors like Arnold Schwarzenegger.

I will focus on the latest issue of the annual Green Film Shooting (Green Film Shooting 2016) magazine and *Eco-Cop* to evaluate how this consultancy perceives its role in managing industry communications. Whereas the majority of the documents from consultancies like the Responsible Media Forum strive to communicate seriousness and inscrutability, featuring logos and graphs on basic templates, the GFS cover goes in another direction. The front cover features Emma Watson and the magazine focuses on film stars and other well-known industry personalities. The document even includes a quotation from Arnold Schwarzenegger:

> The media has a powerful role to play in the fight against climate change. Through films, television, and all media outlets, we must continue to deliver the message that solutions are out there and are happening now. And I believe films in particular can really inspire and make people want to take action. It's great to see some of my film-industry friends working with climate related organizations to push forward those messages.
>
> (ibid.: 3)

The use of celebrity endorsements against glamorous backdrops are typical strategies used by environmental communicators to appeal to wider audiences. They also draw on what Bourdieu conceptualized as symbolic power to facilitate recognition between the communicators and their publics of the importance of the theme communicated. This is significant as it emphasizes the argument that the power to influence rarely relies on forms of coercion but is instead based on mutual value systems. Such a perspective suggests that the best way to get the industry to adopt these measures is by encouraging them to do so, not forcing them by establishing regulation and policies. The use of symbolic elements provides companies like Green Film Shooting with cultural capital that is translated into a variety of ordering discourses in their promotional material. In this case, the idea is to encourage the industry to adapt sustainability measures as the way of the future.

The *Eco-Cop* magazine is a particularly illustrative example of the popular strategies used to entice the industry. The magazine cover is graced by Schwarzenegger superimposed on a glacier and most of its case studies are of international blockbusters such as *Amazing Spider-Man 2* (2014) and the Leonardo DiCaprio film, *The Great Gatsby* (2013). As with the other Green Film Shooting productions, its main focus is on targeting senior management and producers working in the European audio-visual industries. Most of the document focuses on managerial strategies, discussing producers making a commitment to sustainability planning at the earliest stages of the production by communicating with various department heads. Simultaneously, its data provision aims to convey a general level of scientific consensus on the need to adopt green practice in the industry. Key statistics such as the impact of ICT, based on a study performed by the European Broadcast Union (EBU), are complemented by arguments such as 'in France, the audio-visual sector is responsible for discharging approximately one million tons of carbon dioxide into the atmosphere every year' (Green Film Shooting 2015b: 3). To act on these statistics, the document provides interested readers with the opportunity to start implementing green practice with an Excel tool that can calculate a figure based on both cost considerations and carbon emissions. While the discussion includes in-depth exploration of innovations taking place in the industry, such as Myco Foam (an alternative to Styrofoam) and sustainably produced gowns for premieres, it summarizes its aim to make green production easier and more accessible and to also create excitement over these developments instead of obfuscating imageries. Through these mechanisms, Green Film Shooting provides new ordering discourses that may have a considerable impact on integrating environmental sustainability and film production.

Conclusion: the case for film policy

These European case studies provide some solutions to the problems identified with the Nordic film industry. The detrimental role of responsibility deficits, the restrictions on budgets, the reliance on international productions, the absence of any real understanding of sector-specificity and a lack of coordination within and among the film institutes are all addressed here. The arguments for budgeting for sustainability and establishing training and protocols are all significant, but perhaps the most intriguing come from collaborations on sector-specificity. Ecoprod is leading the work here as a particularly important OPP for European production as it was conceptualized as 'a network of film-funding agencies, broadcasters, and environmental agencies ... inspired by the Producers Guild of America's Unified Best Practice Guide' (Green Film Shooting 2015b: 2). Olivier-René Veillon, the chairman of Ecoprod, suggests there is no excuse not to go green as 'more and more producers, especially American studios, are going to be asking for a production's carbon footprint, and they're going to calculate it' (ibid.: 7). These discussions also counteract some of the concerns of the Copenhagen Film Fund on complexity and time management – concerns that apply to the industry as a whole – as they suggest that 'This isn't going to cause the producer any trouble, and it's not going to create any difficulties because calculating the carbon footprint of a production is easy,' state Gammeltoft and Gjerulff, in an interview in 2016.

The importance of working on developing calculators for the sector and for the region as a whole is noted by Lucia Grenna and Donald Ranvaud from Cine-regio:

> We feel it is possible to establish universal precepts for use as best practice guidelines in every country and I think it's inevitable that we will eventually be able to establish a universal, standard protocol. To film financing institutions, we're proposing the idea that the last 10 percent of either the tax incentives or financial support be tied to a production, reducing its carbon footprint. We need to get the whole film industry behind this in a big way.
>
> (Green Film Shooting 2015b: 20)

We have seen examples of this with the policies established by the Flanders Film Fund, the Netherlands Film Fund, Ecoprod, and Schleswig-Holstein in Germany. All of these organizations come to the table with their own variations on similar policies, yet, in Sweden, Film i Skåne talks about the need to adopt standards to the climate in Southern Sweden and to have a national policy instead of a set of regional ones. The correlative work of Birgit Heidsiek and Green Film Shooting brings these together with Cine-regio to emphasize that not only does there need be more understanding of what sustainability means for film and television as a whole, but that we need to account for diversity in unity with comprehensive national film policies to consolidate the discrepant approaches necessitated by specific regulatory regimes in each country.

It is clear from our analysis that the film and television industry needs its own set of strategies and unique protocols to account for its particular needs. We have seen BAFTA engage in extensive training sessions and Richard Reitinger, Head of the Hamburg Media School (HMS), has started to develop educational initiatives to enhance the field via the 'cross-pollination' of students from environmental science with film production. These could lead to further work with the industry: 'We are offering specific work-shops for producers, production managers, camera crews, production designers, and catering teams' (Green Film Shooting 2016: 5), says Christiane Dopp from HMS. But all these developments rely on political support and increasingly from allocation of funds from contested cultural and environmental budgets.

The industry is in the early stages of incorporating environmental practice along its supply and production chain. These are based on vertical negotiations between legislators and industry as well as horizontal cooperation between individual producers and consultancies in different parts of Europe. The outcome is that environmental sustainability is being normalized through repetition of the positive impacts that these strategies can have. While green strategies are still considered an 'obligation' and as an 'obstacle', hesitation is mitigated by reframing them through a variety of mechanisms, including legislative and economic terms. The framing of the debates as a way for companies to cut down on their resource use and expenditure makes for a good argument but it in some ways also contradicts the environmental rationale for adopting them. It is clear that adopting these strategies is, in reality, a balancing act which requires certain rhetorical adjustments, whether to do with cost savings or the ease with which they can be achieved. Yet, the reliance on the rhetoric of sustainable development raises significant questions over the effectiveness of these policies. Most Hollywood productions are able to use a range of means to offset the emissions they generate including recruiting staff or even compensating for them by purchasing carbon offsets. But they also raise questions over how effective these strategies are for total carbon management. After all, if these are only about capital-intensive real-locations of responsibility, nothing will change in practice and the emissions status quo remains. For the production ecologies of the Nordic countries, and for most of Europe, these would not work. Consolidating policy for both institutional governance as well as for production management inside media organizations is, once again, essential as a first step towards an ecological habitus for the media.

Bibliography

BAFTA. 2014. *Albert – Programmes that Don't Cost the Earth*. London: BAFTA.

Cine-regio. 2014. *Sustainability in Vision: Emerging Film and Television Practices and Methodologies in Europe's Regions*. Brussels: Cine-regio.

Green Film Shooting. 2015a. 'A new way of thinking'. Available at: http://greenfilmshooting.net/blog/en/2015/10/29/a-new-way-of-thinking/ (accessed 11 November 2017).

Green Film Shooting. 2015b. *ECO-COP*. Hamburg: Green Film Shooting.

Green Film Shooting. 2016. *Special – Berlinale*. Hamburg: Green Film Shooting.

Greening Film. 2014. 'BS 8909 guidance notes'. Available at: www.greeningfilm.com/what-can-i-do/bs-8909-requirements/bs-8909-guidance-notes-final-4 (accessed 11 November 2017).

Greening Film. 2017. 'A new standard for sustainability on film'. Available at: www.greeningfilm.com/ (accessed 11 November 2017).

Hjort, Mette and Petrie, Duncan. 2007. *The Cinema of Small Nations*. Edinburgh: Edinburgh University Press.

Vaughan, Hunter. 2016. *Sustainable Media*. New York: Routledge, pp. 23–37.

10 Conclusion

Balancing between the footprint and the brainprint

Introduction

> Cinema is and has always been environmentally determined and environmentally determining.
>
> (Bozak 2011: 4)

Bozak's definition of the ways the cinematic apparatus both relies on natural resources and contributes to popular awareness of environmental issues gets to the heart of the debate on the relationship between media and the environment. This book has outlined some of the contemporary strategies that different media sectors – broadcasting, publishing, film and television – have for evaluating and negating their material impact. Yet, many of the organizations evaluated here have suggested that content, not context, can create the largest environmental impact. These arguments focus on the brainprint of media content on audiences, perhaps creating an increased sense of awareness or even changing their opinions on environmental issues. Christian Toennesen, for example, has argued that he would much rather see a 'media company working actively on its brainprint than operational footprint. The impacts there are so much bigger than in changing a few light bulbs' (pers. comm., July 2015).

Is the emphasis on footprinting the media industry a sensible approach or would these energies be better used for working on more effective means of conceptualizing the impact of content? Focusing on the creation of environmental content and targeting behavioural transformation make a lot of sense considering the KPIs of the industry – the principle is to generate content in an economically sustainable way. At the same time, the Responsible Media Forum has emphasized that the sector considers environmental matters an operational aspect of their performance. This roughly translates to circumstances where environmental considerations are seen as useful for reputation management and increasing the company profile, but not in providing improvements on economic or legal performance. Environmental issues can, in some instances, graduate to material concerns and impact the economic operations of a company, such as when a publication develops a reputation for environmental reporting (such as with *The Guardian*) or a film breaks out at the box office (*An Inconvenient Truth*). These successes are rare and do not provide an economically sustainable strategy adaptable across the industry. Whereas for heavy industries, environmental concerns may influence their financial or legal status, the core competencies of media production simply do not leave such a huge production footprint so as to influence their financial or legal performance – at least for now.

Yet, establishing any form of policy to promote environmental content is a very difficult task complicated by a range of factors. Challenges come from areas like the arm's length principle in the Nordic countries which prohibits cultural institutions from any explicit encouragement or influence over content, including environmental programming. Simultaneously, leaving the content shaping the brainprint to develop according to the guiding hand of the free market does not seem like a very solid strategy either, especially in terms of the ethical dimensions of environmental arguments. Private companies can conduct content management much more efficiently than publicly funded companies, but prioritizing environmental content may require data to suggest that such content can be economically profitable in the long run. A much more realistic approach to generating environmental content is to place elements into narratives or backgrounds of sets (such as the use of recycling bins in *EastEnders*).

While BAFTA is now considering extending Albert to cover content, it is difficult to conceptualize how this would take shape in practice. At the same time, the impact of the media's footprint is now being taken seriously by many parts of the industry who are working on their own strategies. This is where policy on material impacts could play a key role. If legislating for content faces considerable hurdles and cannot ever be a truly comprehensive or quantifiable measure, a focus on footprint provides an efficient means to install a solid level of control on the industry's environmental impact. Policy on both measuring and reporting on the footprint requires the industry to focus its gaze on its own actions and provide both quantifiable analytical data on its emissions and explain these measures in qualifiable ways to both their staff and their stakeholders. While imposed top-down control is never the most efficient management tool, especially on a topic as personal as environmental issues, it does indicate a viable way to introduce environmental sustainability to production practice. Supporting this is the fact that the industry is at a very early stage of adoption, and the implications of mechanisms like the contracts for the BBC's supply chain or the Flanders Film Fund requiring a CO_2 report for the last 10 per cent of funds are significant in ensuring these practices are taken seriously. The aim is to encourage the industry through soft power strategies to accept sustainability as an everyday part of production. Simultaneously, establishing these strategies for both the sector as a whole and through specific media institutions could provide a pragmatic way to avoid the responsibility deficit.

Yet, the exact content and directions of such policies are still open to debate. Currently, fragmentation is one of the key obstacles facing the industry. This goes for both geopolitical alignment and sectoral coordination. The EU has a range of codes and directives that govern the infrastructure of the media. The EU Environmental Code on both resource management and recycling influences some areas of operations as do the WEEE and RoHS. National governments have strict rules on emissions and in most cases have aspirations to go CO_2-neutral. All these areas influence the infrastructure of the media and ensure that large aspects of their operations are sustainable. While such areas comprise a huge part of the emissions of the operations of media companies, more specific regulation on emissions is required. Part of the rationale is that sectoral differences act as one of the obstacles to developing functional policy for the media as a whole. A better understanding of the distinctive strategies required to meet each part of the industry appropriately would move the debate forward considerably. Thus, there would need to be correlation between the regulatory and operational environments of each sector. Here, publishing tends to focus on print, which comes with a visible and widely acknowledged material footprint. Broadcasting sees most of its emissions from travel and large-scale operational areas like offices and studios. Film tends to have most of its emissions generated by location shooting, but also deals with general concerns for the industry such as building management, set construction, accommodation, digital infrastructure, and so on. While all these areas have different priorities, there is clear overlap between them to provide the building blocks for the development of solid media policy that would also be able to take into account the particular core competencies of each media form.

The development of environmental sustainability initiatives in the Nordic media industries highlights both challenges and potential solutions to these ongoing problems. By the time of the publication of this book, many of the initiatives described here will have been superseded by both regulatory and technological developments. It would seem we are now on the cusp of environmental protocols finally becoming much more standardized and normalized practices in the industry. Once this is reality, the analysis of environmental sustainability will need a whole new set of methods to evaluate these protocols outside of the more discursive approach taken here. Another factor that will have repercussions across the media discussed here is the role of digital communications in both production and dissemination of content. Areas such as the amount of emissions from streaming content to using pervasive media devices are only now beginning to be understood by climate scientists. As we have suggested, the emissions generated by accessing and transferring data on mobile devices exceeds emissions from the hardware by a considerable margin. Similarly, the significance of the renewable infrastructure will transform with, for example, the introduction of large-scale server farms in Finland and Sweden by Google and Facebook to make use of the cool climate. The launch of domestic data centre operators such as GreenQloud and Green Mountain in Iceland and Norway will influence the total media footprint of these countries. There is thus an urgent need to move from the first phase of analysing organizational management to production management. Understanding sustainability as a network of different agents – regulators, companies, producers, crew, consumers, consultancies, institutes, and a whole range of other stakeholders, as well as the constitutive role of the material base of all these operations – would ensure that what we have referred to as grounded materialities gain an even more prominent role as KPIs of organizational strategy for media companies.

Bibliography

Bozak, Nadia. 2011. *The Cinematic Footprint: Lights, Camera, Environment.* New Brunswick, NJ: Rutgers University Press.

Index

Milton Keynes UK
Ingram Content Group UK Ltd.
UKHW040103071024
449327UK00019B/769